AF291856

Springer Proceedings in Mathematics & Statistics

Volume 123

Springer Proceedings in Mathematics & Statistics

This book series features volumes composed of selected contributions from workshops and conferences in all areas of current research in mathematics and statistics, including operation research and optimization. In addition to an overall evaluation of the interest, scientific quality, and timeliness of each proposal at the hands of the publisher, individual contributions are all refereed to the high quality standards of leading journals in the field. Thus, this series provides the research community with well-edited, authoritative reports on developments in the most exciting areas of mathematical and statistical research today.

More information about this series at http://www.springer.com/series/10533

Ingrid Bauer · Shelly Garion
Alina Vdovina
Editors

Beauville Surfaces and Groups

 Springer

Editors
Ingrid Bauer
Lehrstuhl Mathematik VIII
University of Bayreuth
Bayreuth
Germany

Shelly Garion
Institut für Mathematische Logik und
  Grundlagenforschung
University of Münster
Münster
Germany

Alina Vdovina
School of Mathematics and Statistics
Newcastle University
Newcastle-upon-Tyne
UK

ISSN 2194-1009   ISSN 2194-1017   (electronic)
Springer Proceedings in Mathematics & Statistics
ISBN 978-3-319-13861-9   ISBN 978-3-319-13862-6   (eBook)
DOI 10.1007/978-3-319-13862-6

Library of Congress Control Number: 2015933367

Mathematics Subject Classification: 11 Number theory (11E57, 11G32), 14 Algebraic Geometry
(14J10, 14J29, 14J50, 14H30, 14L30), 20 Group Theory and Generalizations (20B25, 20D06, 20D15,
20E42, 20F29, 20G40)

Springer Cham Heidelberg New York Dordrecht London

© Springer International Publishing Switzerland 2015
This work is subject to copyright. All rights are reserved by the Publisher, whether the whole or part
of the material is concerned, specifically the rights of translation, reprinting, reuse of illustrations,
recitation, broadcasting, reproduction on microfilms or in any other physical way, and transmission
or information storage and retrieval, electronic adaptation, computer software, or by similar or
dissimilar methodology now known or hereafter developed.
The use of general descriptive names, registered names, trademarks, service marks, etc. in this
publication does not imply, even in the absence of a specific statement, that such names are exempt
from the relevant protective laws and regulations and therefore free for general use.
The publisher, the authors and the editors are safe to assume that the advice and information in this
book are believed to be true and accurate at the date of publication. Neither the publisher nor the
authors or the editors give a warranty, express or implied, with respect to the material contained
herein or for any errors or omissions that may have been made.

Printed on acid-free paper

Springer International Publishing AG Switzerland is part of Springer Science+Business Media
(www.springer.com)

Contents

Introduction

Ingrid Bauer, Shelly Garion and Alina Vdovina

"Beauville Surfaces" are certain rigid regular surfaces of general type, which can be described in a purely algebraic combinatorial way. Thus they play a very important role in different fields of mathematics such as algebraic geometry, group theory and number theory. The notion of Beauville surface was introduced by Fabrizio Catanese in 2000 and after the first systematic study of these surfaces by Ingrid Bauer, Fabrizio Catanese and Fritz Grunewald, there has been an increasing interest in the subject.

The conference "Beauville Surfaces and Groups 2012" was held in the University of Newcastle, UK, from 7 to 9 June 2012. This conference brought together, for the first time, experts from different fields of mathematics interested in Beauville surfaces, as well as young researchers and Ph.D. students, from the UK, USA, Germany, Italy and Spain, in order to share the status of the art and to discuss further developments in the study of these fascinating surfaces.

These conference proceedings include 11 chapters discussing Beauville surfaces (and generalizations of them), from different points of view, algebro-geometric and group-theoretic. Some of the chapters are expository survey papers and the others are research chapters on recent developments in this area. The chapters reflect the topics of the lectures presented during the workshop, and discuss various open problems and conjectures related to Beauville surfaces.

We briefly describe the content of the chapters.

The first chapter, by Ingrid Bauer, Fabrizio Catanese and Davide Frapporti, gives a comprehensive overview of what is currently known about the fundamental groups and, in particular, about the first homology groups of Beauville surfaces, and more generally, surfaces isogenous to a higher product of curves. The authors provide a computer script which calculates the first homology groups of regular surfaces of general type, hence in particular for Beauville surfaces.

The chapter by Nathan Barker, Nigel Boston, Norbert Peyemirnhoff and Alina Vdovina presents new interesting examples of finite groups admitting unmixed ramification structures giving rise to regular algebraic surfaces isogenous to a higher product of curves, generalizing Beauville surfaces. These groups are constructed as certain p-quotients of particular groups with special presentation related to finite projective planes and to expander graphs.

Nigel Boston overviews the important role played by a very interesting class of groups, namely p-groups, in the study of Beauville surfaces. The chapter describes recent results as to which p-groups admit a Beauville structure, with emphasis on ones of small order (due to Barker, Boston and Fairbairn) and ones that form inverse systems (due to Barker, Boston, Peyerimhoff and Vdovina).

Ben Fairbairn gives new interesting examples of finite groups which define real Beauville surfaces, and in particular, new examples of infinite families of groups admitting strongly real Beauville structures. The chapter discusses certain finite simple groups as well as abelian and nilpotent groups. It moreover deals with characteristically simple groups and almost simple groups.

Shelly Garion surveys the probabilistic group-theoretical approach towards proving well-known conjectures of Bauer, Catanese and Grunewald regarding Beauville surfaces arising from finite simple groups, by describing the following three works. The first is the work of Garion, Larsen and Lubotzky, showing that almost all finite simple groups of Lie type admit a Beauville structure. The second is the work of Garion and Penegini on Beauville structures of alternating groups, based on results of Liebeck and Shalev, and the third is the case of the group $PSL_2(q)$, in which the author gives bounds on the probability of generating a Beauville structure.

The contribution of Christian Gleissner gives a complete classification of the connected components of the moduli space of surfaces of general type corresponding to regular unmixed surfaces S isogenous to a product of curves with $\chi(S) = 2$. These surfaces are natural generalizations of Beauville surfaces and the results require similar techniques from computational group theory. The surfaces studied in this research paper are no longer rigid, whence they come up in families, which are completely described.

Gareth Jones presents new interesting infinite families of Beauville surfaces arising from characteristically simple groups. More precisely, the author proves that if G is a cartesian power of a finite simple group of Lie type of low Lie rank or a sporadic simple group, then G is a Beauville group if and only if it has two generators and is not isomorphic to A_5. The author moreover conjectures that any cartesian power of a finite simple group of Lie type, which is 2-generated and not isomorphic to A_5, admits a Beauville structure.

The chapter by Kay Magaard and Chris Parker discusses the natural question of which extensions of the finite simple groups admit Beauville structures. This follows the previous exciting results due to Guralncik and Malle and to Fairbairn, Maggard and Parker, that all finite simple (respectively, quasi-simple) groups (except A_5 and $SL_2(5)$) admit Beauville structures, thus proving a well-known conjecture of Bauer, Catanese and Grunewald. The chapter gives various interesting new examples of Frattini covers and semidirect products of finite simple groups admitting Beauville structures. As an application the authors deduce that every finite quotient of the profinite group $SL_d(\mathbb{Z}_p)$, where $d \geq 9$, admits a Beauville structure.

Matteo Penegini gives asymptotic estimates on the number of connected components of the moduli space of surfaces of general type, focusing on interesting families of surfaces isogenous to a higher product of curves, and Beauville surfaces in particular, and describes recent results due to Garion and Penegini. The proof is based on group theoretical methods, using the group theoretical terminology of the so-called "ramification structures", and Beauville structures in particular.

The paper by Roberto Pignatelli surveys the state of the art on quasi-étale quotients of a product of two curves, i.e. surfaces which are the quotient of a product of two curves by a finite group, acting freely outside a finite set of points. This class of surfaces has been successfully used in the last years to construct many interesting examples of surfaces of general type in a systematic computer aided way. The chapter reviews the principal results and gives a complete list of minimal quasi-étale surfaces of general type with geometric genus equal to the irregularity ≤ 2.

Francesco Polizzi gives a survey on recent work on isotrivially fibred surfaces and their numerical invariants. One of the main goals is, if $\alpha : X \to C$ is a relatively minimal isotrivial fibration of the surface X over a curve C, to relate the invariants K_X^2 and $\chi(\mathcal{O}_X)$ by some inequalities. Since surfaces X admitting an isotrivial fibration are of the form $X = C_1 \times C_2/G$, where $C_i, i = 1, 2$ are curves of respective genera at least two, and G is a finite group acting faithfully (but not necessarily freely) on $C_1 \times C_2$, these surfaces are natural generalizations of Beauville surfaces.

The editors would like to thank all the authors contributing to this volume and the referees for their assistance. The editors are grateful to Springer Proceedings in Mathematics and Statistics (PROMS) for publication of this collection of chapters.

The editors are grateful to all the speakers and participants in the workshop "Beauville Surfaces and Groups 2012", to the University of Newcastle for hosting the workshop, and to Nathan Barker for his assistance in organizing the workshop. The editors want to thank the London Mathematical Society (LMS) and the Deutsche Forschungsgemeinschaft (DFG) for their financial support.

These conference proceedings are dedicated to the memory of Fritz Grunewald, *who was the first to realize how fascinating Beauville surfaces are and inspired many mathematicians from different areas to investigate these surfaces. He is deeply missed.*

The Fundamental Group and Torsion Group of Beauville Surfaces

Ingrid Bauer, Fabrizio Catanese and Davide Frapporti

Abstract We give a survey on the fundamental group of surfaces isogenous to a higher product. If the surfaces are regular, e.g. if they are Beauville surfaces, the first homology group is a finite group. We present a MAGMA script which calculates the first homology groups of regular surfaces isogenous to a product.

1 Introduction

One aim of this note is to give an overview of what is known about the fundamental groups and more specifically about the first homology groups of Beauville surfaces and of their generalizations. In particular, we wrote a MAGMA script which calculates the torsion group of Beauville surfaces, or more generally of regular surfaces isogenous to a higher product.

Fundamental groups of algebraic varieties are a very interesting area of research and they are still quite mysterious objects, since usually it may be very hard to determine them. On the other hand, if one can calculate them in some specific cases (e.g., for surfaces of general type) they are a powerful tool to conclude that two surfaces (of general type) are not contained in the same connected component of their moduli space.

Here we will restrict ourselves to the following situation: let C_1 and C_2 be projective algebraic curves of respective genera g_1, g_2 at least 2 and let G be a subgroup of the automorphism group $\mathrm{Aut}(C_1 \times C_2)$. We denote by X the quotient $(C_1 \times C_2)/G$; when X is singular, we denote by S the minimal resolution of singularities of X.

We define $G^0 := G \cap (\mathrm{Aut}(C_1) \times \mathrm{Aut}(C_2))$. Then (compare [10] for this and the following assertions) G^0 is a subgroup of index at most 2 inside $G < \mathrm{Aut}(C_1 \times C_2)$ and acts on each factor and "diagonally" on the product $C_1 \times C_2$ (i.e., for each $\gamma \in G^0$, we can write $\gamma = (\gamma_1, \gamma_2)$, and $\gamma(x_1, x_2) = (\gamma_1(x_1), \gamma_2(x_2))$).

I. Bauer (✉) · F. Catanese · D. Frapporti
Lehrstuhl Mathematik VIII, Mathematisches Institut der Universität Bayreuth, NW II,
Universitätsstr. 30, 95447 Bayreuth, Germany
e-mail: Ingrid.bauer@uni-bayreuth.de

© Springer International Publishing Switzerland 2015
I. Bauer et al. (eds.), *Beauville Surfaces and Groups*, Springer Proceedings
in Mathematics & Statistics 123, DOI 10.1007/978-3-319-13862-6_1

By [10], it is always possible to assume that G^0 acts faithfully on both factors: in this case we say that $(C_1 \times C_2)/G$ is a *minimal realization*, and this minimal realization is moreover unique. From now on we shall assume that we have indeed a minimal realization. Then (see [10]) there are two cases.

The *mixed case* is the case where the action of G exchanges the two factors: in this case $C_1 \cong C_2$ and $G^0 \triangleleft_2 G$.

The *unmixed case* is the case where $G = G^0$, and its two projections into $\mathrm{Aut}(C_j)$, $j = 1, 2$, are injective. Identifying then G to a subgroup $G \subset \mathrm{Aut}(C_j)$, for $j = 1, 2$, we say that G acts diagonally: this means that we view G as the diagonal subgroup inside $G \times G \subset (\mathrm{Aut}(C_1) \times \mathrm{Aut}(C_2))$.

In the unmixed case, S is called a *product-quotient surface* and X is its *quotient model*; while in the mixed case, S is a *mixed surface* and X is a *mixed quotient*.

In this note we want to focus on the *unmixed case*, so from now on we implicitly assume that $G = G^0$. Some of the results can be extended/generalized to the mixed case, but the approach is slightly different. We refer to [13, 14] for further details.

In the first section we describe product-quotient surfaces and the algebraic data which determine them. In the second section we comment on the results on the fundamental group of product-quotient surfaces and their higher-dimensional analogues. The third section is dedicated to the description of a MAGMA [8] script which, given as input a finite group G and two ordered tuples

$$T_1 = (n_1, \ldots, n_r), \quad T_2 = (m_1, \ldots, m_s),$$

has as output

- one representative surface $S = (C_1 \times C_2)/G$ for each irreducible connected component of the moduli space, where G is acting freely with signatures (T_1, T_2),
- for each such representative, the first homology group $H_1(S, \mathbb{Z})$ of S.

Finally, in the last section, we show several concrete and explicit calculations, obtained through a direct application of our program.

2 Product-Quotient Surfaces

Let G be a finite group and let S be a *product-quotient surface* with *quotient model* $X = (C_1 \times C_2)/G$. If G acts freely, then X is smooth, and we have the following:

Definition 2.1 A surface S is said to be *isogenous to a (higher) product* if S is the quotient $(C_1 \times C_2)/G$, where $g_i := genus(C_i) \geq 2$, and G is a finite group acting freely on $C_1 \times C_2$.

The word "higher" means that the respective genera of C_1, C_2 are ≥ 2, in particular this implies that S is of general type (and with ample canonical divisor). However, for commodity, from now on we shall drop the word "higher".

In the last years, a huge amount of new surfaces of general type with $p_g = q$ have been constructed as the quotient of a product of two curves by the action of a finite group; see [1, 3, 4, 6, 7, 13, 14] for $p_g = 0$, [11, 13, 17, 19, 20] for $p_g = 1$, [18, 23] for $p_g = 2$.

In particular, we have a complete classification of surfaces isogenous to a product with $p_g = q$. The case where G does not act freely is still open (and particularly difficult in the regular case $q = 0$), in spite of several results. A difficulty which is peculiar of the regular case is that the following can happen: the minimal resolution S of singularities X may not be minimal surface. Up to now there are almost no techniques to decide whether there are exceptional curves of the first kind (i.e., smooth rational curves with self intersection (-1)) on S and how to find them explicitly (cf. [7]).

The description of product-quotient surfaces is accomplished through the theory of Galois coverings between projective curves (also named "Riemann surfaces").

Definition 2.2 Let $g \geq 0$ and $m_1, \ldots, m_r > 1$ be integers. The *orbifold surface group of signature* $(g; m_1, \ldots, m_r)$ is defined as:

$$\mathbb{T}(g; m_1, \ldots, m_r) := \langle a_1, b_1, \ldots, a_g, b_g, c_1, \ldots, c_r \mid$$
$$c_1^{m_1}, \ldots, c_r^{m_r}, \prod_{i=1}^{g}[a_i, b_i] \cdot c_1 \cdots c_r \rangle.$$

Given a finite group H, a *generating vector* for H of signature $(g; m_1, \ldots, m_r)$ is a $(2g + r)$-tuple of elements of H:

$$V := (d_1, e_1, \ldots, d_g, e_g; h_1, \ldots, h_r)$$

such that

- V generates H,
- $\prod_{i=1}^{g}[d_i, e_i] \cdot h_1 \cdot h_2 \cdots h_r = 1$ and
- there exists a permutation $\sigma \in \mathfrak{S}_r$ such that $\mathrm{ord}(h_i) = m_{\sigma(i)}$ for $i = 1, \ldots, r$.

If $g = 0$, then $V := (h_1, \ldots, h_r)$ is called a *spherical system of generators* of H of signature $(m_1, \ldots, m_r)$.

To give a generating vector of signature $(g; m_1, \ldots, m_r)$ for a finite group H is equivalent to give an *appropriate orbifold homomorphism*

$$\psi \colon \mathbb{T}(g; m_1, \ldots, m_r) \longrightarrow H,$$

i.e., a surjective homomorphism ψ such that $\psi(c_i)$ has order m_i.

Remark 2.3 By Riemann's existence theorem (see [5]), any curve C of genus g together with an action of a finite group H on it, such that C/H is a curve C' of genus g', is determined (modulo automorphisms) by the following data:

(1) the branch point set $\{p_1, \ldots, p_r\} \subset C'$;

(2) generators $\alpha_1, \ldots, \alpha_{g'}, \beta_1, \ldots, \beta_{g'}, \gamma_1, \ldots, \gamma_r$ of $\pi_1(C' \setminus \{p_1, \ldots, p_r\})$, where each γ_i is a simple geometric loop around p_i and

$$\prod_{i=1}^{g'} [\alpha_i, \beta_i] \cdot \gamma_1 \cdots \cdot \gamma_r = 1;$$

(3) a generating vector V for H of signature $(g'; m_1, \ldots, m_r)$ with the property that *Hurwitz's formula* holds:

$$2g - 2 = |H| \left(2g' - 2 + \sum_{i=1}^{r} \frac{m_i - 1}{m_i} \right). \tag{2.1}$$

Remark 2.4 Let C be a curve with an action of a finite group H on it and consider the Galois covering $c \colon C \to C/H$. The appropriate orbifold homomorphism $\psi \colon \mathbb{T}(g'; m_1, \ldots, m_r) \to H$ associated to V is induced by the monodromy of the Galois étale H-covering $c^0 \colon C^0 \to C'^0$ induced by c, where C'^0 is the curve obtained from C' by removing the branch points $p_1, \ldots, p_r$ of c, and $C^0 := c^{-1}(C'^0)$.

Let $h_i := \psi(c_i) \in H$: h_i is called the *local monodromy element* and generates the stabilizer of a point in $c^{-1}(p_i)$. We also define

$$\Sigma(V) := \bigcup_{g \in G} \bigcup_{j \in \mathbb{Z}} \bigcup_{i=1}^{r} \{ g \cdot h_i^j \cdot g^{-1} \},$$

as the union of the stabilizers for the action of H on C.

Remark 2.5 Let $S \to X := (C_1 \times C_2)/G$ be a product-quotient surface. The two Galois coverings $C_i \to C_i/G =: C_i'$ determine a pair (V_1, V_2) of generating vectors of G. The action of G on $C_1 \times C_2$ is free if and only if (V_1, V_2) is *disjoint*, that is

$$\Sigma(V_1) \cap \Sigma(V_2) = \{1\}.$$

Conversely, a pair (V_1, V_2) of generating vectors of G of respective signatures $T_1 = (g_1'; m_1, \ldots, m_r)$ and $T_2 = (g_2'; n_1, \ldots, n_s)$, determines a family of product-quotient surfaces of dimension $M_1 + r + M_2 + s$, where $M_i = 3g_i' - 3$.

In particular, the product-quotient surface can only be rigid (i.e., has no non-trivial deformations), if $r = s = 3$ and $g_1' = g_2' = 0$. If the action is non free, then it can happen that the above family has dimension 0, still the surface is not rigid, i.e., it has non trivial deformations which are no longer product-quotient surfaces.

Instead, if S is isogenous to a product, it follows by the result of the second author below that S is rigid if and only if $r = s = 3$ and $g_1' = g_2' = 0$.

Theorem 2.6 (see [10]) *(a) A projective smooth surface S is isogenous to a product of two curves of respective genera $g_1, g_2 \geq 2$, if and only if the following two conditions are satisfied:*

(1) there is an exact sequence

$$1 \to \Pi_{g_1} \times \Pi_{g_2} \to \pi = \pi_1(S) \to G \to 1,$$

where G is a finite group and where Π_{g_i} denotes the fundamental group of a projective curve of genus $g_i \geq 2$;

(2) $e(S)(= c_2(S)) = \dfrac{4}{|G|}(g_1 - 1)(g_2 - 1)$.

(b) Write $S = (C_1 \times C_2)/G$. Any surface X with the same topological Euler number and the same fundamental group as S is diffeomorphic to S and is also isogenous to a product. There is a smooth proper family with connected smooth base manifold T, $p \colon \mathcal{X} \to T$ having two fibres respectively isomorphic to X, and Y, where Y is one of the four surfaces $S = (C_1 \times C_2)/G$, $S_{+-} := (\overline{C_1} \times C_2)/G$, $\overline{S} = (\overline{C_1} \times \overline{C_2})/G$, $S_{-+} := (C_1 \times \overline{C_2})/G = \overline{S_{+-}}$.

(c) The corresponding subset of the moduli space of surfaces of general type $\mathfrak{M}_S^{top} = \mathfrak{M}_S^{diff}$, corresponding to surfaces orientedly homeomorphic, resp. orientedly diffeomorphic to S, is either irreducible and connected or it contains two connected components which are exchanged by complex conjugation.

In particular, if S' is orientedly diffeomorphic to S, then S' is deformation equivalent to S or to $\overline{S}$.

Definition 2.7 Let $S = (C_1 \times C_2)/G$ be a surface isogenous to a product. S is called a *Beauville surface* if $r = s = 3$ and $g(C_1/G) = g(C_2/G) = 0$.

A corollary of the above Theorem 2.6 (see [10]) is that a Beauville surface is rigid, it has no nontrivial deformations.

Remark 2.8 Observe that, by [9] (cf. also [21]), if $S \to X := (C_1 \times C_2)/G$ is a product-quotient surface, then $q(S) = g(C_1/G) + g(C_2/G)$. Therefore, any Beauville surface is regular.

The concept of Beauville surfaces was introduced by the second author in 1997, and their global rigidity was shown in [10]. They were first systematically studied in [2], where the connection between the algebro-geometric background and the group theoretic description was explained in more detail, many new examples were constructed, and a lot of conjectures were presented. Quite a number of these conjectures have been solved in the meantime and there exists a substantial literature on Beauville surfaces nowadays, which is reflected by the several contributions to the present volume.

3 The Fundamental Group

Let $S := (C_1 \times C_2)/G$ be the minimal realization of a surface isogenous to a product: then the surface $C_1 \times C_2$ is a Galois étale covering space of S with group G and we have already observed (in Theorem 2.6) that the fundamental group of S sits into an exact sequence

$$1 \longrightarrow \pi_1(C_1) \times \pi_1(C_2) \longrightarrow \pi_1(S) \longrightarrow G \longrightarrow 1.$$

If one drops the assumption about the freeness of the action of G on the product $C_1 \times C_2$, there is no reason that the behaviour of the fundamental group of the quotient should be similar to the above situation. Nevertheless, in the unmixed case, it turns out that the fundamental group $\pi_1(X)$ admits a very similar description.

Definition 3.1 We call the fundamental group $\Pi_g := \pi_1(C)$ of a projective curve of genus g a *(genus g) surface group*.

Theorem 3.2 (see [4]) *Let C_1 and C_2 be projective curves of genus at least 2 and let G be a finite group acting on each C_i as a group of automorphisms and diagonally on the product $C_1 \times C_2$.*

Let S be the minimal desingularization of $X := (C_1 \times C_2)/G$. Then the fundamental group $\pi_1(X) \cong \pi_1(S)$ has a normal subgroup $\mathcal{N}$ of finite index which is isomorphic to the product of surface groups, i.e., there are integers $h_1, h_2 \geq 0$ such that $\mathcal{N} \cong \Pi_{h_1} \times \Pi_{h_2}$.

Remark 3.3 (1) The previous theorem holds also in dimension $n \geq 3$, see [4, 12].
(2) In the case of surfaces isogenous to a product we have that h_i equals the genus of C_i, for $i = 1, 2$.

Especially the case of infinite fundamental groups is the one where the above structure theorem turns out to be extremely helpful in order to give an explicit description of these groups. As we will explain in the sequel, it is easy to get a presentation of the fundamental group of a product-quotient surface. But, in general, a presentation of a group does not say much about the group. Even the problem whether it is trivial or not is in general an undecidable problem. Here, since we know that π_1 is a "surface times surface by finite" group, we go through its normal subgroups of finite index until such a subgroup $\mathcal{N}$ appears.

We recall now how to compute the fundamental group of a product-quotient surface $S \to X = (C_1 \times C_2)/G$ starting from the associated algebraic data:

$$\psi_i : \mathbb{T}_i \longrightarrow G \qquad i = 1, 2.$$

Remark 3.4 The kernel of ψ_i is isomorphic to the fundamental group $\pi_1(C_i)$, and the action of $\pi_1(C_i)$ on the universal cover $u : \hat{C}_i \to C_i$ extends to a properly discontinuous action of $\mathbb{T}_i$. Moreover, u is ψ_i-equivariant and $\hat{C}_i/\mathbb{T}_i \cong C_i/G_i$.

There are two short exact sequences:

$$1 \longrightarrow \pi_1(C_i) \longrightarrow \mathbb{T}_i \xrightarrow{\psi} G \longrightarrow 1, \quad i = 1, 2. \tag{3.1}$$

We define the fibre product

$$\mathbb{H} := \mathbb{H}(G; \psi_1, \psi_2) := \{(x, y) \in \mathbb{T}_1 \times \mathbb{T}_2 \mid \psi_1(x) = \psi_2(y)\}.$$

The exact sequences in (3.1) induce then an exact sequence:

$$1 \to \pi_1(C_1) \times \pi_1(C_2) \to \mathbb{H}(G; \psi_1, \psi_2) \to \Delta_G \cong G \to 1,$$

where $\Delta_G \subset G \times G$ denotes the diagonal subgroup.

Definition 3.5 Let H be a group. Then $\mathrm{Tors}(H)$ is the normal subgroup generated by the torsion elements of H (i.e., the elements of finite order in H).

Proposition 3.6 ([4, Proposition 3.4]) *Let $S \to X = (C_1 \times C_2)/G$ be a product-quotient surface and let*

$$\psi_i : \mathbb{T}_i \longrightarrow G \quad i = 1, 2$$

be the associated appropriate orbifold homomorphisms. Then

$$\pi_1(S) = \pi_1(X) = \mathbb{H}/\mathrm{Tors}(\mathbb{H}).$$

A very special and theoretically easy case is:

Corollary 3.7 *Let $S = (C_1 \times C_2)/G$ be a surface isogenous to a product. Then $\pi_1(S) = \mathbb{H}$ and $H_1(S, \mathbb{Z}) = \mathbb{H}^{ab}$.*

In the literature there are several articles on the construction and classification of surfaces birational to the quotient of a product of curves; we have already cited most of them. In some cases the authors provide a computer-script to compute the fundamental group (or the first homology group) of the surfaces they study.

Nevertheless, in this note we take the opportunity to give a simple algorithm in the special case, which includes the case Beauville surfaces, where $S = (C_1 \times C_2)/G$ is a *regular* surface isogenous to a product (equivalently, $g'_1 = g'_2 = 0$).

We now recall how to determine whether two regular surfaces isogenous to a product of unmixed type are deformation equivalent (cf. [1, 3]).

By Remark 2.8, and as observed above, S regular means that $C_i/G \cong \mathbb{P}^1, i = 1, 2$. We denote by $\mathcal{B}(G; T_1, T_2)$ the set of disjoint pairs (V_1, V_2) of spherical systems of generators of G of respective signatures T_1 and T_2.

Let $V := [h_1, \ldots, h_r]$ be a r-tuple of elements of G and $1 \leq i \leq r$. We consider the usual *Hurwitz move* $\sigma_i(V)$ defined by

$$\sigma_i(V) := [h_1, \ldots, h_{i-1}, h_i h_{i+1} h_i^{-1}, h_i, h_{i+2}, \ldots, h_r].$$

It is well known that $\sigma_1, \ldots, \sigma_r$ generate the braid group on r letters $\mathbf{B}_r$, and that $\mathbf{B}_r$ maps spherical system of generators to spherical system of generators, and preserving the signature. Also the automorphism group $\text{Aut}(G)$ of G acts on the set of spherical systems of generators of a fixed signature by simultaneous application of an automorphism φ to the coordinates of a r-tuple: $\varphi(V) := [\varphi(h_1), \ldots, \varphi(h_r)]$. We get the following action of $\mathbf{B}_r \times \mathbf{B}_s \times \text{Aut}(G)$ on $\mathcal{B}(G; T_1, T_2)$: let $(\gamma_1, \gamma_2, \varphi) \in \mathbf{B}_r \times \mathbf{B}_s \times \text{Aut}(G)$ and $(V_1, V_2) \in \mathcal{B}(G; T_1, T_2)$, where T_1 has length r and T_2 has length s, we set

$$(\gamma_1, \gamma_2, \varphi) \cdot (V_1, V_2) := (\varphi(\gamma_1(V_1)), \varphi(\gamma_2(V_2))).$$

We denote this action by $\mathcal{H}$.

Theorem 3.8 (cf. [3, Theorem 5.2]) *Let S and S' be two regular surfaces, both isogenous to a product of unmixed type, and with associate pairs of spherical system of generators (V_1, V_2), respectively (V_1', V_2').*

Then S and S' are deformation-equivalent if and only if the respective groups are isomorphic, $G(S) \cong G(S')$, and either (V_1, V_2) and (V_1', V_2') belong to the same $\mathcal{H}$-orbit or (V_1, V_2) and (V_2', V_1') do.

Remark 3.9 Given a finite group G and two signatures T_1 and T_2 of length r and s, it is then work for a computer to determine $\mathcal{B}(G; T_1, T_2)$, its orbits for the $\mathcal{H}$-action and the first homolgy group $H_1(S, \mathbb{Z})$ of the associate surfaces S.

We have written a MAGMA [8] script (see Appendix section "Appendix: The Script") which takes as input G, T_1 and T_2, and returns as output a representative $(V_1, V_2) \in \mathcal{B}(G; T_1, T_2)$ for each $\mathcal{H}$-orbit, together with the first homology group $H_1(S, \mathbb{Z})$ of the associate surface S. The outline of the script is the following:

- Step 1: we compute all the spherical systems of generators of G of respective signatures T_1 and T_2 and we collect them in orbits for the action of the braid group $\mathbf{B}_r$ (resp. $\mathbf{B}_s$).
- Step 2: we discard the pairs of orbits of non-disjoint spherical systems of generators.
- Step 3: the remaining pairs yield surfaces isogenous to a product. We consider the action of $Aut(G)$ on them, hence we get the $\mathcal{H}$-orbits in $\mathcal{B}(G; T_1, T_2)$. Observe that indeed it suffices to consider only the action of $Out(G)$, since the $Inn(G)$-action was already taken care of in the first step.
- Step 4: we run over the outputs of Step 3 and we compute their first homology group $H_1(S, \mathbb{Z})$.

4 Some Applications

Surfaces isogenous to a product of unmixed type with $p_g = 0$ have been classified in [3]. We run our script for these surfaces and we get:

Table 1 Surfaces isogenous to a product of unmixed type with $p_g = q = 0$

G	$Id(G)$	T_1	T_2	N	D	$H_1(S, \mathbb{Z})$
$\mathcal{A}_5$	$\langle 60, 5 \rangle$	$[2, 5, 5]$	$[3, 3, 3, 3]$	1	1	$(\mathbb{Z}_3)^2 \times (\mathbb{Z}_{15})$
$\mathcal{A}_5$	$\langle 60, 5 \rangle$	$[5, 5, 5]$	$[2, 2, 2, 3]$	1	1	$(\mathbb{Z}_{10})^2$
$\mathcal{A}_5$	$\langle 60, 5 \rangle$	$[3, 3, 5]$	$[2, 2, 2, 2, 2]$	1	2	$(\mathbb{Z}_2)^3 \times \mathbb{Z}_6$
$\mathcal{S}_4 \times \mathbb{Z}_2$	$\langle 48, 48 \rangle$	$[2, 4, 6]$	$[2, 2, 2, 2, 2, 2]$	1	3	$(\mathbb{Z}_2)^4 \times \mathbb{Z}_4$
$G(32)$	$\langle 32, 27 \rangle$	$[2, 2, 4, 4]$	$[2, 2, 2, 4]$	1	2	$(\mathbb{Z}_2)^2 \times \mathbb{Z}_4 \times \mathbb{Z}_8$
$(\mathbb{Z}_5)^2$	$\langle 25, 2 \rangle$	$[5, 5, 5]$	$[5, 5, 5]$	1	0	$(\mathbb{Z}_5)^3$
$\mathcal{S}_4$	$\langle 24, 12 \rangle$	$[3, 4, 4]$	$[2, 2, 2, 2, 2, 2]$	1	3	$(\mathbb{Z}_2)^4 \times \mathbb{Z}_8$
$G(16)$	$\langle 16, 3 \rangle$	$[2, 2, 4, 4]$	$[2, 2, 4, 4]$	1	2	$(\mathbb{Z}_2)^2 \times \mathbb{Z}_4 \times \mathbb{Z}_8$
$D_4 \times \mathbb{Z}_2$	$\langle 16, 11 \rangle$	$[2, 2, 2, 4]$	$[2, 2, 2, 2, 2, 2]$	1	4	$(\mathbb{Z}_2)^3 \times (\mathbb{Z}_4)^2$
$(\mathbb{Z}_2)^4$	$\langle 16, 14 \rangle$	$[2, 2, 2, 2, 2]$	$[2, 2, 2, 2, 2]$	1	4	$(\mathbb{Z}_4)^4$
$(\mathbb{Z}_3)^2$	$\langle 9, 2 \rangle$	$[3, 3, 3, 3]$	$[3, 3, 3, 3]$	1	2	$(\mathbb{Z}_3)^5$
$(\mathbb{Z}_2)^3$	$\langle 8, 5 \rangle$	$[2, 2, 2, 2, 2]$	$[2, 2, 2, 2, 2, 2]$	1	5	$(\mathbb{Z}_2)^4 \times (\mathbb{Z}_4)^2$

Theorem 4.1 *Let $S = (C_1 \times C_2)/G$ be a surface isogenous to a product of unmixed type, with $p_g(S) = 0$, then G is one of the groups in the Table 1 and the signatures are listed in the table. The number N of components in the moduli space, their dimension D and the first homology group of S are given in the last three columns.*

Remark 4.2 (1) In a previous paper [1] (containing the classification of surfaces isogenous to a product of unmixed type with $p_g = q = 0$ and G abelian) there is a mistake, an erroneous statement about commutators in a product. The error resulted into finding only a proper quotient of the actual first homology group.

(2) The correct calculation of the first homology group was done more than one year ago by the first author, who used a MAGMA script to find the correct answer.

(3) Another correction to do to [1] is that the authors forgot the possibility of swapping factors, hence there is only one isomorphism class of surfaces in the case $G = (\mathbb{Z}_5)^2$.

(4) Our computer calculations confirm the result of the paper [22], dedicated to the calculation of the first homology groups of surfaces isogenous to a product of unmixed type with $p_g = q = 0$ and G abelian.

In [16], the author classifies the regular surfaces isogenuos to a product of unmixed type with $\chi(\mathcal{O}) = 2$; in particular he classifies the unmixed Beauville surfaces with $p_g = 1$. Applying our program we get the following:

Theorem 4.3 (cf. [16]) *Let $S = (C_1 \times C_2)/G$ be a Beauville surface of unmixed type, with $p_g(S) = 1$. Then G is one of the groups in Table 2 and the signatures are listed in the table. The number N of components in the moduli space and the first homology group of S are given in the last two columns.*

Table 2 Unmixed Beauville surfaces with $p_g = 1$

G	$Id(G)$	T_1	T_2	N	$H_1(S, \mathbb{Z})$
$PSL(2, 7) \times \mathbb{Z}_2$	$\langle 336, 209 \rangle$	$[2, 3, 14]$	$[4, 4, 4]$	2	$(\mathbb{Z}_4)^2$
$PSL(2, 7)$	$\langle 168, 42 \rangle$	$[7, 7, 7]$	$[3, 3, 4]$	2	$\mathbb{Z}_7 \times \mathbb{Z}_{21}$
$PSL(2, 7)$	$\langle 168, 42 \rangle$	$[3, 3, 7]$	$[4, 4, 4]$	2	$\mathbb{Z}_4 \times \mathbb{Z}_{12}$
$G(128, 36)$	$\langle 128, 36 \rangle$	$[4, 4, 4]$	$[4, 4, 4]$	2	$(\mathbb{Z}_2)^3 \times (\mathbb{Z}_4)^2$

Table 3 Some unmixed Beauville surfaces with group $G = PSL(2, q)$

G	$Id(G)$	T_1	T_2	N	$H_1(S, \mathbb{Z})$	$\chi(\mathcal{O}_S)$
$PSL(2, 7)$	$\langle 168, 42 \rangle$	$[3, 3, 4]$	$[7, 7, 7]$	2	$\mathbb{Z}_7 \times \mathbb{Z}_{21}$	2
		$[3, 4, 4]$	$[7, 7, 7]$	1	$\mathbb{Z}_7 \times \mathbb{Z}_{28}$	4
		$[4, 4, 4]$	$[7, 7, 7]$	2	$(\mathbb{Z}_{28})^2$	6
		$[4, 4, 4]$	$[3, 7, 7]$	4	$\mathbb{Z}_4 \times \mathbb{Z}_{28}$	4
		$[4, 4, 4]$	$[3, 3, 7]$	2	$\mathbb{Z}_4 \times \mathbb{Z}_{12}$	2
$PSL(2, 8)$	$\langle 504, 156 \rangle$	$[2, 7, 7]$	$[3, 3, 9]$	3	$\mathbb{Z}_3 \times \mathbb{Z}_{21}$	6
		$[2, 7, 7]$	$[3, 9, 9]$	3	$\mathbb{Z}_3 \times \mathbb{Z}_{63}$	12
		$[2, 7, 7]$	$[9, 9, 9]$	7	$\mathbb{Z}_9 \times \mathbb{Z}_{63}$	18
		$[7, 7, 7]$	$[9, 9, 9]$	14	$(\mathbb{Z}_{63})^2$	48
		$[7, 7, 7]$	$[2, 9, 9]$	6	$\mathbb{Z}_7 \times \mathbb{Z}_{63}$	20
		$[7, 7, 7]$	$[2, 3, 9]$	6	$\mathbb{Z}_7 \times \mathbb{Z}_{21}$	4
		$[7, 7, 7]$	$[3, 9, 9]$	6	$\mathbb{Z}_{21} \times \mathbb{Z}_{63}$	32
		$[7, 7, 7]$	$[3, 3, 9]$	6	$(\mathbb{Z}_{21})^2$	16
$PSL(2, 9) \cong \mathcal{A}_6$	$\langle 360, 118 \rangle$	$[3, 3, 5]$	$[4, 4, 4]$	1	$(\mathbb{Z}_{12})^2$	3
		$[3, 5, 5]$	$[4, 4, 4]$	1	$\mathbb{Z}_4 \times \mathbb{Z}_{60}$	6
		$[5, 5, 5]$	$[4, 4, 4]$	2	$\mathbb{Z}_{20} \times \mathbb{Z}_{60}$	9
				2	$(\mathbb{Z}_{20})^2$	
		$[5, 5, 5]$	$[3, 3, 5]$	2	$\mathbb{Z}_5 \times \mathbb{Z}_{15}$	3
				2	$(\mathbb{Z}_{15})^2$	

Let $G = \mathrm{PSL}(2, q)$, where q and is a prime power and $q \leq 9$. For $q \leq 5$, no group has a disjoint pair of spherical generators (see [2]), while in the other cases the "admissible" pairs of signatures are classified (see [15]). The outputs of our script are collected in Table 3. We use the same notation of the previous tables and we add a column reporting the Euler characteristic $\chi(\mathcal{O}_S)$.

In Table 4, we collect the outputs in some other easy cases: $G \in \{\mathcal{S}_5, \mathcal{S}_6, (\mathbb{Z}_7)^2\}$. One can prove (using a simple MAGMA script), that if $X = (C_1 \times C_2)/G$ is a Beauville surface with $G \in \{\mathcal{S}_5, \mathcal{S}_6, (\mathbb{Z}_7)^2\}$, then its signature (T_1, T_2) is in Table 4.

Table 4 Other unmixed Beauville surfaces with a given group

G	$Id(G)$	T_1	T_2	N	$H_1(S, \mathbb{Z})$	$\chi(\mathcal{O}_S)$
$\mathcal{S}_5$	$\langle 120, 34 \rangle$	$[4, 4, 5]$	$[3, 6, 6]$	1	$\mathbb{Z}_3 \times \mathbb{Z}_{24}$	3
$\mathcal{S}_6$	$\langle 720, 763 \rangle$	$[5, 6, 6],$	$[4, 6, 6]$	8	$(\mathbb{Z}_6)^2$	35
		$[5, 6, 6],$	$[4, 4, 6]$	16	$\mathbb{Z}_2 \times \mathbb{Z}_{24}$	28
		$[5, 6, 6],$	$[4, 4, 4]$	8	$\mathbb{Z}_4 \times \mathbb{Z}_{12}$	21
		$[3, 6, 6],$	$[4, 4, 4]$	5	$(\mathbb{Z}_{12})^2$	15
		$[2, 5, 6],$	$[4, 4, 4]$	1	$(\mathbb{Z}_4)^2$	6
$(\mathbb{Z}_7)^2$	$\langle 49, 2 \rangle$	$[7, 7, 7]$	$[7, 7, 7]$	7	$(\mathbb{Z}_7)^3$	4

Appendix: The Script

```
// TuplesOfGivenOrder creates a sequence of the same length as the input
// sequence type, whose entries are subsets of the group in the input,
// and precisely the subsets of elements of order the corresponding
// entry of type

TuplesOfGivenOrders:=function(G,type)
SEQ:=[];
for i in [1..#type-1] do
  EL:={g: g in G| Order(g) eq type[i]};
  if IsEmpty(EL) then return [{}];
    else Append(~SEQ,EL);
end if; end for;
return SEQ;
end function;

// This function transforms a tuple into a sequence.

TupleToSeq:=func<tuple|[x: x in tuple]>;

// Now we create all sets of spherical generators of a group of the
// prescribed signature.

VectGens:=function(G, type)
Vect:={}; SetCands:=TuplesOfGivenOrders(G,type);
for cand in CartesianProduct(SetCands) do
  if Order(&*cand) eq type[#type] then
    if #sub<G|TupleToSeq(cand)> eq #G then
      Include(~Vect, Append(TupleToSeq(cand),(&*cand)^-1));
end if; end if; end for;
return Vect;
end function;

// HurwitzOrbit, starting from a sequence seq of elements of a group,
// creates all sequences of elements which are equivalent to the given one
// for the equivalence relation generated by the Hurwitz moves
// and returns (to spare memory) only the ones whose entries have
// orders disposed as the ones in seq.
```

```
HurwitzMove:= func<seq,j|Insert(Remove(seq,j),j+1,seq[j]^seq[j+1])>;

HurwitzOrbit:=function(seq)
orb:={ }; shortorb:={  }; Trash:={ seq};
repeat ExtractRep(~Trash,~gens); Include(~orb, gens);
  for k in [1..#seq-1] do
    newgens:=HurwitzMove(gens,k);
    if newgens notin orb then Include(~Trash, newgens);
end if; end for; until IsEmpty(Trash);
for gens in orb do test:=true;
  for k in [1..#seq] do
    if not Order(gens[k]) eq Order(seq[k]) then
      test:=false; break k;
  end if; end for;
  if test then Include(~shortorb, gens); end if;
end for;
return shortorb;
end function;


// OrbitsVectGens creates a sequence
// whose entries are subsets of the group in the input,
// and precisely the subsets  are the orbits under the
// Hurwitz moves.

OrbitsVectGens:=function(G, type)
Orbits:=[];   Vects:=VectGens(G, type);
while not IsEmpty(Vects) do
  v:=Rep(Vects);   orb:=HurwitzOrbit(v);
  Append(~Orbits, orb); Vects:=Vects diff orb;
end while;
return Orbits;
end function;


// Disjoint checks if two spherical systems of generators are disjoint.

Stab:= function(seq, G)
M:= {} ;
for i in [1..#seq] do g:=seq[i];
  for n in [1 .. (Order(g)-1) ] do
    M := M join Conjugates(G,g^n) ;
end for; end for;
return M;
end function ;

Disjoint:=func<G,seq1, seq2 | IsEmpty( Stab(seq1,G) meet Stab(seq2,G))>;

// Homology computes the first homology group
// of the surface associated to the disjoint pair
// of spherical systems of generators (seq1, seq2) of G.

Poly:=function(seq, gr)
F:=FreeGroup(#seq);   Rel:={}; Q:=Id(F);
for i in {1..#seq} do
  Q:=Q*F.(i); Include(~Rel,F.(i)^(Order(seq[i])));
end for;
Include(~Rel,Q); P:=quo<F|Rel>;
return P, hom<P->gr|seq>;
end function;
```

```
Homology:=function(G, seq1,  seq2)
T1,f1:=Poly(seq1,G);        T2,f2:=Poly(seq2,G);
T1xT2:=DirectProduct(T1,T2);
 GxG,inG:=DirectProduct(G,G);
if Category(G) eq GrpPC then
  n:=NPCgens(G);
  else n:=NumberOfGenerators(G);
end if;
Diag:= hom<G->GxG| [inG[1](G.j)*inG[2](G.j): j in [1..n]]>(G);
f:=hom<T1xT2->GxG| inG[1](seq1) cat inG[2](seq2)>;
Pi1:=Rewrite(T1xT2,Diag@@f);
return AbelianQuotient(Pi1);
end function;


// Surfaces is the main function.
// It takes as input a group G and two signatures.
// It calls the previous function and moreover identifies
// distinct pairs of Hurwitz's orbits of spherical systems of
// generators under the action of Aut(G).
// Note that we consider only the action of Out(G), since
// the Inn(G)-action was already taken care of in HurwitzOrbit.


Surfaces:=function(G, type1,type2)
R:={}; Aut:=AutomorphismGroup(G);
F,q:=FPGroup(Aut);       O1,p:=OuterFPGroup(Aut);
Out,k:=PermutationGroup(O1);
V1:=OrbitsVectGens(G, type1);
if type1 eq type2 then V2:=V1;
  else V2:=OrbitsVectGens(G,type2);
end if;
W:=Set(CartesianProduct({1..#V1},{1..#V2}));
for pair  in W do
  v1:= Rep(V1[pair[1]]); v2:= Rep(V2[pair[2]]);
  if not Disjoint(G, v1,v2) then Exclude(~W, pair);
 end if;
end for;
while not IsEmpty(W) do
  pair:=Rep(W);
  v1:= Rep(V1[pair[1]]); v2:= Rep(V2[pair[2]]);Include(~R,[v1,v2]);
  for phi in Out do
    if exists(x){y: y in W | (q(phi@@(p*k)))(v1) in V1[y[1]] and
     (q(phi@@(p*k)))(v2) in V2[y[2]]} then Exclude(~W, x);
      if type1 eq type2 then Exclude(~W, <x[2],x[1]>);
    end if; end if;
    if IsEmpty(W) then break phi; end if;
  end for; end while;
printf "Number of components: %o\n", #R;
for r in R do
  printf "Spherical generators:\n%o\n%o\n", r[1],r[2];
  printf "Homology:\n%o\n", Homology(G,r[1],r[2]);
end for;
return {};
end function;
```

References

1. I. Bauer, Some new surfaces with $p_g = q = 0$, in *The Fano Conference*, Turin University, Torino, pp. 123–142 (2004)
2. I. Bauer, F. Catanese, F. Grunewald, Beauville surfaces without real structures, *Geometric Methods in Algebra and Number Theory*, volume 235 of Progress in Mathematics (Birkhäuser, Boston, 2005), pp. 1–42
3. I. Bauer, F. Catanese, F. Grunewald, The classification of surfaces with $p_g = q = 0$ isogenous to a product of curves. Pure Appl. Math. Q. **4**(2), 547–586 (2008)
4. I. Bauer, F. Catanese, F. Grunewald, R. Pignatelli, Quotients of products of curves, new surfaces with $p_g = 0$ and their fundamental groups. Am. J. Math. **134**(4), 993–1049 (2012)
5. I. Bauer, F. Catanese, R. Pignatelli, Surfaces of general type with geometric genus zero: a survey, in *Complex and Differential Geometry*. Springer Proceedings in Mathematics, vol. 8 (Springer, Berlin, 2011), pp. 1–48
6. I. Bauer, R. Pignatelli, The classification of minimal product-quotient surfaces with $p_g = 0$. Math. Comput. **81**(280), 2389–2418 (2012)
7. I. Bauer, R. Pignatelli, Product-quotient surfaces: new invariants and algorithms (2013). arXiv:1308.5508
8. W. Bosma, J. Cannon, C. Playoust, The Magma algebra system I. The user language. J. Symbolic Comput. **24**(3–4), 235–265 (1997)
9. F. Catanese, Everywhere nonreduced moduli spaces. Invent. Math. **98**(2), 293–310 (1989)
10. F. Catanese, Fibred surfaces, varieties isogenous to a product and related moduli spaces. Am. J. Math. **122**(1), 1–44 (2000)
11. G. Carnovale, F. Polizzi, The classification of surfaces with $p_g = q = 1$ isogenous to a product of curves. Adv. Geom. **9**(2), 233–256 (2009)
12. T. Dedieu, F. Perroni, The fundamental group of a quotient of a product of curves. J. Group Theory **15**(3), 439–453 (2012)
13. D. Frapporti, R. Pignatelli, Mixed quasi-étale quotients with arbitrary singularities. Glasg. Math. J. **57**(1), 143–165 (2015)
14. D. Frapporti, Mixed quasi-étale surfaces, new surfaces of general type with $p_g = 0$ and their fundamental group. Collect. Math. **64**(3), 293–311 (2013)
15. S. Garion, On Beauville structures for PSL (2,q) (2010). arXiv:1003.2792
16. C. Gleißner, in *The classification of regular surfaces isogenous to a product of curves with* $\chi(\mathcal{O}_S) = 2$. ed. by I. Bauer, S. Garion, A. Vdovina. Beauville Surfaces and Groups (Springer, Cham, 2015)
17. E. Mistretta, F. Polizzi, Standard isotrivial fibrations with $p_g = q = 1$ II. J. Pure Appl. Algebra **214**(4), 344–369 (2010)
18. M. Penegini, The classification of isotrivially fibred surfaces with $p_g = q = 2$. Collect. Math. **62**(3), 239–274 (2011) (With an appendix by Sönke Rollenske)
19. F. Polizzi, On surfaces of general type with $p_g = q = 1$ isogenous to a product of curves. Commun. Algebra **36**(6), 2023–2053 (2008)
20. F. Polizzi, Standard isotrivial fibrations with $p_g = q = 1$. J. Algebra **321**(6), 1600–1631 (2009)
21. F. Serrano, Isotrivial fibred surfaces. Ann. Mat. Pura Appl. **171**(4), 63–81 (1996)
22. T.I. Shabalin, Homology of some surfaces with $p_g = q = 0$ isogenous to a product. Izv. RAN. Ser. Mat. **78**(6), 211–221 (2014)
23. F. Zucconi, Surfaces with $p_g = q = 2$ and an irrational pencil. Can. J. Math. **55**(3), 649–672 (2003)

Regular Algebraic Surfaces, Ramification Structures and Projective Planes

N. Barker, N. Boston, N. Peyerimhoff and A. Vdovina

Abstract Regular algebraic surfaces isogenous to a higher product of curves can be obtained from finite groups with ramification structures. We find unmixed ramification structures for finite groups constructed as p-quotients of particular infinite groups with special presentation related to finite projective planes.

2000 Mathematics Subject Classification: 14L30 · 20F32 · 51E24

1 Introduction

An algebraic surface is *isogenous to a higher product* (of curves) if it admits a finite unramified covering which is isomorphic to a product of curves $C_1 \times C_2$ of genera $g(C_i) \geq 2$. It was shown in [10] that every such surface S has a unique *minimal realisation* $S \cong (C_1 \times C_2)/G$, where G is a finite group acting freely on $C_1 \times C_2$ and C_1 and C_2 have the smallest possible genera. Moreover, G respects the product structure by either acting diagonally on each factor (unmixed case) or there are elements in G interchanging the factors (mixed case). Surfaces isogenous to a higher product are always minimal and of general type. Particularly interesting examples

N. Barker (✉) · A. Vdovina
Department for Pure Mathematics and Mathematical Statistics,
University of Cambridge, CB3 OWB, Cambridge, UK
e-mail: nb443@cam.ac.uk

A. Vdovina
e-mail: alina.vdovina@ncl.ac.uk

N. Boston
Department of Mathematics,
University of Wisconsin, 303 Van Vleck Hall, 480 Lincoln Drive,
Madison, WI 53706, USA
e-mail: boston@math.wisc.edu

N. Peyerimhoff
Department of Mathematical Sciences, Science Laboratories,
Durham University, South Road, DH1 3LE, Durham, UK
e-mail: norbert.peyerimhoff@durham.ac.uk

© Springer International Publishing Switzerland 2015
I. Bauer et al. (eds.), *Beauville Surfaces and Groups*, Springer Proceedings
in Mathematics & Statistics 123, DOI 10.1007/978-3-319-13862-6_2

are *Beauville surfaces*, i.e., algebraic surfaces isogeneous to a higher product which are rigid (i.e., do not admit nontrivial deformations).

The *irregularity* $q(S)$ of a surface S is the difference between its geometric and its algebraic genus, and agrees with the Hodge number $h^{1,0}(S)$. Surfaces with vanishing irregularity are called *regular*. Since $q(S) = g(C_1/G) + g(C_2/G)$ (see [24, Proposition 2.2]), we have $C_i/G \cong \mathbb{P}^1$ for both curves in the minimal realisation of a regular surface.

Every surface S isogenous to a higher product gives rise to a finite group G via its minimal realisation. This process can be reversed. Starting with a finite group G, the existence of a so called *ramification structure* can be used to construct a regular surface of the form $(C_1 \times C_2)/G$. We will discuss ramification structures and the construction of the associated surfaces in Sect. 2. Bauer, Catanese and Grunewald [5] used this group theoretical description to classify all regular surfaces S isogenous to a product of curves with vanishing geometric genus $p_g(S) = h^{2,0}(S)$. The process in [5] was aided by the reduction of the search of ramification structures to groups of order less than 2000, for which the MAGMA library of small groups could then be used. They saw this classification as the solution in a very special case to the open problem posed by Mumford: "Can a computer classify all surfaces of general type with $p_g = 0$?"

The infinite group in [15, Example 6.3] given by the presentation

$$G_0 := \langle x_0, \ldots, x_6 \mid x_i x_{i+1} x_{i+3} \ (i \in \mathbb{Z}_7) \rangle \tag{1}$$

was used in [1, 2] to construct finite 2-groups with special unmixed and mixed ramification structures, giving rise to unmixed and mixed Beauville surfaces. These finite 2-groups were the maximal 2-quotients of 2-class k of both the group G_0 and its index two subgroup H_0 generated by x_0 and x_1. We like to mention that very little is known about *mixed* Beauville structures and it is generally assumed that they are very rare. Results in [13, 14] show that the symmetric groups S_n and all almost simple groups with sporadic derived groups cannot have mixed structures. The examples in [2] are the first known *infinite* family of 2-groups with mixed Beauville structures. It is immediate from the definition that a p-group can only admit a mixed structure if $p = 2$. The only other known construction of groups admitting mixed Beauville structures was given in [4]. But this general construction is very different in nature and does not provide examples of 2-*groups* admitting mixed Beauville structures. In this paper we restrict our considerations, however, to the *unmixed case*.

The above group G_0 belongs to a family called *groups with special presentation*. These groups were introduced by Howie [15] and are related to projective planes over finite fields (see Sect. 3 for more details). It was proved in [12] that all groups with special presentation are just infinite (i.e., they are infinite groups all of whose non-trivial normal subgroups have finite index). A natural question arose: *Do any other groups with special presentations give rise to finite groups with particular ramification structures?*

In this article we consider finite index subgroups of the groups listed in [11, Example 3.3], an index 3 subgroup of the following group with special presentation from [15, Example 6.4], the group

$$G := \langle x_0, \ldots, x_{12} \mid x_i^3, x_i x_{i+1} x_{i+4} \ (i \in \mathbb{Z}_{13}) \rangle \tag{2}$$

and the group given in [18, Example 2] constructed from a polyhedral presentation (a generalization of the triangle presentations defined in [8]). We use the computer program MAGMA to search for unmixed ramification structures in maximal p-quotients of p-class k of the above mentioned groups for various primes p. These ramification structures give then rise to particular regular surfaces isogenous to a higher product. Moreover, these results lead to natural conjectures about infinite families of p-groups admitting unmixed ramification structures.

Let us present an example of our results. The subgroup H of G defined in (2), and generated by x_0, x_1, x_2 has index 3. Let $H_{3,k}$ be the maximal 3-quotient of 3-class k. For simplicity, let us denote the elements in $H_{3,k}$ corresponding to x_0, x_1, x_2, again, by x_0, x_1, x_2. Let $y_0 = x_0 x_1^2 x_2^2$, $y_1 = x_0^2 x_1 x_2^2$ and $y_2 = x_1 x_2^{-1} x_2^{x_0}$. Then we have the following result.

Theorem 1.1 *For $k = 2, \ldots, 60$, the groups $H_{3,k}$ are of order 3^{a_k} and admit unmixed ramification structures (T_1, T_2) of type $([3,3,3,3^{d_k}], [3^{b_k}, 3^{b_k}, 3^{b_k}, 3^{b_k}])$, where $T_1 = (x_0, x_1, x_2, (x_0 x_1 x_2)^{-1})$, $T_2 = (y_0, y_1, y_2, (y_0 y_1 y_2)^{-1})$, $b_k = 1 + [\log_3 \frac{3k}{4}]$, $d_k = 1 + [\log_3 k]$, and*

$$a_k = \begin{cases} 8j & \text{if } k = 3j, \\ 8j + 3 & \text{if } k = 3j+1, \\ 8j + 6 & \text{if } k = 3j+2. \end{cases}$$

Here $[x]$ denotes the largest integer $\leq x$.

This result motivates the following conjecture.

Conjecture 1.2 *Let H be the index 3 subgroup of the group (2), generated by x_0, x_1, x_2. Then, for all $k \geq 2$, the maximal 3-quotients $H_{3,k}$ of 3-class k are of order 3^{a_k} and admit unmixed ramification structures (T_1, T_2) of type $([3,3,3,3^{d_k}], [3^{b_k}, 3^{b_k}, 3^{b_k}, 3^{b_k}])$, where a_k, b_k, d_k and T_1 and T_2 are explicitly given in Theorem 1.1.*

A promising approach to prove this conjecture is to employ the matrix representation of (2) in Appendix 2: "Representation for the Group G". (Such a matrix representation was key in [2] to prove that infinitely many 2-quotients of the group (1) admit mixed Beauville structures.)

The article is organised as follows. Section 2 presents fundamental facts about ramification structures and algebraic surfaces. Our results on ramification structures are presented in Sect. 3 below. Appendix 1: "Expanders Associated to the Group G_0" is concerned with the derivation of an explicit matrix representation of the group (2). Finally, Appendix 1: "Expanders Associated to the Group G_0" provides a brief survey about explicit recent expander constructions related to the group (1).

2 Ramification Structures and Associated Surfaces

2.1 Group Theoretical Structures

Following [5] closely, we give the definition of an (unmixed) ramification structure of a finite group G.

An r-tuple $T = [g_1, \ldots, g_r]$ of non-trivial elements of G is called a *spherical system of generators*, if $g_1, \ldots, g_r$ generate G and $g_1 g_2 \cdots g_r = 1$. The r-tuple $[m_1, \ldots, m_r]$ of non-decreasing orders of the elements g_i is called the *type* of the spherical system T of generators, i.e., $2 \leq m_1 \leq m_2 \leq \cdots \leq m_r$ and there is a permutation $\tau \in Sym(r)$ such that $m_i = \mathrm{ord}(g_{\tau(i)})$. Let

$$\Sigma(T) := \bigcup_{g \in G} \bigcup_{j=0}^{\infty} \bigcup_{i=1}^{r} \{g g_i^j g^{-1}\}$$

be the union of all conjugates of the cyclic subgroups generated by the elements g_i of the spherical system. Two spherical systems of generators $T_1 = [g_1, \ldots, g_r]$ and $T_2 = [g_1', \ldots, g_s']$ are called *disjoint* if $\Sigma(T_1) \cap \Sigma(T_2) = \{1\}$. An unmixed ramification structure is defined as follows.

Definition 2.1 *(Unmixed ramification structures, see* [5, Definition 1.1]*)* Let $A_1 = [m_1, \ldots, m_r]$ and $A_2 = [n_1, \ldots, n_s]$ be tuples of natural numbers with $2 \leq m_1 \leq \cdots \leq m_r$ and $2 \leq n_1 \leq \cdots \leq n_s$. An *unmixed ramification structure of type* (A_1, A_2) for a finite group G is a pair (T_1, T_2) of disjoint spherical systems of generators such that T_1 has type A_1 and T_2 has type A_2.

The disjointness of the pair (T_1, T_2) of an unmixed ramification structure guarantees that G acts freely on the product $C_1 \times C_2$ of associated algebraic curves (see Sect. 2.2 and the references therein). In this article we will only consider unmixed ramification structures and their associated surfaces. For examples of the mixed case see, e.g., [1, 4, 5].

Recall that *unmixed Beauville structures* are unmixed ramification structures with two spherical systems (T_1, T_2) of length 3, i.e., $r = s = 3$. They are of particular interest, since they give rise to the Beauville surfaces mentioned in the Introduction.

2.2 From Ramification Structures to Algebraic Surfaces

In this section we explain how to construct an algebraic surface $S = (C_{T_1} \times C_{T_2})/G$ from a given finite group G with an unmixed ramification structure (T_1, T_2).

Let G be a finite group and $T = [g_1, \ldots, g_r]$ be a spherical system of generators with $m_i = \mathrm{ord}(g_{\tau(i)})$. For $1 \leq i \leq r$, let $P_1, \ldots, P_l \in \mathbb{P}^1$ be a sequence of points ordered counterclockwise around a base point P_0 and $\gamma_i \in \pi(\mathbb{P}^1 - \{P_1, \ldots, P_r\}, P_0)$ be represented by a simple counterclockwise loop around P_i, such that $\gamma_1\gamma_2\ldots\gamma_r = 1$. By Riemann's existence theorem, we obtain a surjective homomorphism

$$\Phi : \pi(\mathbb{P}^1 - \{P_1,\ldots,P_r\}, P_0) \to G$$

with $\Phi(\gamma_i) = g_i$ and a Galois covering $\lambda : C_T \to \mathbb{P}^1$ with ramification indices equal to the orders of the elements $g_1,\ldots,g_r$. These data induce a well defined action of G on the curve C_T, and by the Riemann-Hurwitz formula, we have

$$g(C_T) = 1 + \frac{|G|}{2}\left(r - 2 - \sum_{l=1}^{r}\frac{1}{m_l}\right). \tag{3}$$

Now, we assume that G admits an unmixed ramification structure (T_1, T_2). This leads to a diagonal action of G on the product $C_{T_1} \times C_{T_2}$, and the disjointness of the two spherical systems of generators ensures that G acts freely on the product of curves. The associated algebraic surface S is the quotient $(C_{T_1} \times C_{T_2})/G$. By the Theorem of Zeuthen-Segre, we have for the topological Euler number

$$e(S) = 4\frac{(g(C_{T_1}) - 1)(g(C_{T_2}) - 1)}{|G|},$$

as well as the relations (see [10, Theorem 3.4]),

$$\chi(S) = \frac{e(S)}{4} = \frac{K_S^2}{8},$$

where K_S^2 is the self intersection number of the canonical divisor and $\chi(S) = 1 + p_g(S) - q(S)$ is the holomorphic Euler-Poincaré characteristic of S. Assume that (T_1, T_2) is of the type (A_1, A_2) with $A_1 = [m_1,\ldots,m_r]$ and $A_2 = [n_1,\ldots,n_s]$. Then the above relations imply for the associated surface S that

$$\chi(S) = \frac{|G|}{4}\left(r - 2 - \sum_{l=1}^{r}\frac{1}{m_l}\right)\left(s - 2 - \sum_{l=1}^{s}\frac{1}{n_l}\right).$$

3 Groups with Special Presentations

As mentioned in [11], small cancellation groups are generalizations of surface groups and satisfy many of the nice properties of those groups. It was proved in [11] (with a small list of exceptions) that almost all groups with a presentation satisfying the small cancellation conditions $C(3)$ and $T(6)$ contain a free subgroup of rank 2.

Further, [15] proved that most $C(3)$, $T(6)$ groups G (namely, the ones which do *not* have special presentations) are SQ-universal. (A group G is called SQ-universal if every countable group can be embedded in a quotient group of G.) In that article, a group presentation was called *special* if every relator has length 3 and the star graph is isomorphic to the incidence graph of a finite projective plane. (See [21, p. 61] for a textbook reference on the *star graph* of a presentation. In short, the star graph of a presentation $\langle \mathbf{x} \mid \mathbf{r} \rangle$ is defined as follows (see [15]): its vertex set are the elements in $\mathbf{x} \cup \mathbf{x}^{-1}$ and there is an edge from x to y if there is a cyclically reduced word beginning in x and ending in y, which is a cyclic permutation of a relator or its inverse.) Moreover, it was asked (see [15, Question 6.11]) whether any or all of the groups with special presentations are SQ-universal. It was proved in [12] that groups with special presentation are just infinite (i.e., all non-trivial normal subgroups have finite index) and, therefore, cannot be SQ-universal. (Note that special presentations in the sense of [15] are $(3, 3)$-special in the sense of [12].)

Howie [15] also set up an example machine (see Theorem 3.1 below) to create infinitely many groups with special presentations. More precisely, he constructed a special presentation with star graph isomorphic to the incidence graph of the projective plane over every finite field $\mathbb{F}_q$, where q is a prime power. Until then, only seven examples of special presentations were known (see [11, Example 3.3]), and each of them has a star graph isomorphic to the Heawood graph (i.e., the incidence graph of the 7-point projective plane over $\mathbb{F}_2$).

Given a finite field $K = \mathbb{F}_q$ (for q a prime power), a *positive* presentation with star graph isomorphic to this incidence graph of the Desarguesian projective plane over K is formed.

The construction takes a cubic extension of K, namely $F = \mathbb{F}_{q^3}$, and identifies the cyclic group $C_m = F^\times / K^\times$ with the points of the projective pane $\mathcal{P}$ over K, where $m = q^2 + q + 1$.

The group C_m acts on $\mathcal{P}$ via multiplication in F, and this action is regular on both the points and lines of $\mathcal{P}$ i.e. C_m is a Singer group, see [16]. The lines of $\mathcal{P}$ can be identified with the subset σL of C_m, where σ ranges over C_m and L is a fixed line or perfect difference set i.e. a set of residues $a_1, \ldots, a_{q+1} \mod m$ such that every non-zero residue modulo $m = q^2 + q + 1$ can be expressed uniquely in the form $a_i - a_j$.

Theorem 3.1 ([15, Theorem 6.2]) *Let q be a prime power and $m = q^2 + q + 1$. Then there exists a subset l of $q+1$ elements of $\mathbb{Z}_m$ such that*

$$\langle x_0, \ldots, x_{m-1} \mid x_i x_{i+\lambda} x_{i+\lambda+q\lambda} \ (i \in \mathbb{Z}_m, \lambda \in l) \rangle$$

is a special presentation whose star graph is isomorphic to the incidence graph of the projective plane over $GF(q)$.

Let us now present results on ramification structures of finite groups obtained from particular groups G with special presentations. These finite groups are generated via the lower, exponent p-central series, i.e.,

$$G = P_0(G) \geq \cdots \geq P_{i-1}(G) \geq P_i(G) \geq \ldots,$$

where $P_i(G) = [P_{i-1}(G), G]P_{i-1}(G)^p$ for $i \geq 1$. The finite groups $G_{p,k}$ under considerations are then the maximal p-quotients of p-class k, denoted by $G_{p,k}$ and given by $G_{p,k} = G/P_k(G)$.

The results discussed below are obtained via the computer program MAGMA (see [6]). Note that the algorithm `pQuotient` constructs, for a given group G, a consistent power-conjugate presentation for $G_{p,k}$.

3.1 Ramification Structures for the Group in [15, Example 6.3]

The group G_0 in (1) with seven generators $x_0, \ldots, x_6$ appeared as Example 6.3 in [15]. (G_0 is constructed using Theorem 3.1 with $q = 2$, $m = 7$ and $l = \{1, 2, 4\}$.) The subgroup H_0 generated by x_0, x_1 has index two. In [1, Theorems 4.1 and 4.2], we presented unmixed ramification structures for the 2-groups $(H_0)_{2,k}$ for $3 \leq k \leq 64$ (for k not a power of 2), as well as mixed ramification structures for the 2-groups $(G_0)_{2,k}$ for $3 \leq k \leq 10$ (again, for k not a power of 2). Since the involved spherical systems of generators consist of three elements, these ramification structures are actually Beauville structures and lead to new examples of Beauville surfaces.

Moreover, [2] presents a rigorous proof that an *infinite family* of 2-quotients of G_0 admits mixed Beauville structures. [2] uses a faithful matrix representation of G_0 by infinite upper triangular matrices (created in an analogous way as described in Appendix 1: "Expanders Associated to the Group G_0"), and the quotients are obtained via truncations at upper diagonals. MAGMA calculations show that the first 100 of these quotients are isomorphic to the quotients $(G_0)_{2,k}$. It is natural to conjecture that all of these quotients of G_0 constructed in these two different ways are pairwise isomorphic (see [22, Conjecture 1]).

The group G_0 appears also in [9, Sect. 4] as the group A.1. (The articles [8, 9] are concerned with simply transitive group actions on the vertices of $\widetilde{A}_2$-buildings.) The index two subgroup H_0 was also used in [17, 22] to construct families of expander graphs of vertex degrees 4 and 3. These expander graph constructions are briefly described in Appendix 1: "Expanders Associated to the Group G_0".

3.2 Ramification Structures for the Groups
in [11, Example 3.3]

There, a list of seven special group presentations $G_i = \langle \mathbf{x} \mid \mathbf{r}_i \rangle$, $1 \leq i \leq 7$, with

$$r_1 = \{ab^{-1}d, a^{-1}dc, d^{-1}ea, b^2f, ceg, cgf, efg\},$$
$$r_2 = \{a^{-1}df, b^{-1}ed, c^{-1}fe, a^2g, bdg, bec, cgf\},$$
$$r_3 = \{abc, ade, afg, cge, bef, bdg, dfc\},$$
$$r_4 = \{a^2b, acd, bde, bfc, ceg, dgf, efg\},$$

$$r_5 = \{a^2b, acd, bde, bfc, ceg, dg^2, ef^2\},$$
$$r_6 = \{a^2b, acb, bef, bge, d^2f, c^2g, egf\},$$
$$r_7 = \{abc, adb, acd, bef, cge, dfg, egf\},$$

were given ($\mathbf{x} = \{a, b, c, d, e, f, g\}$). The star graphs of all seven presentations are isomorphic to the incidence graph of the 7-point projective plane. See Fig. 1 for the star graph of the group G_1.

Our group G_0 in (1) coincides with their group G_3, which was discussed in Sect. 3.1. It is stated in [11] that the only isomorphism between abelianised groups G_i^{ab} is between G_4^{ab} and G_6^{ab}. However, if one looks at the commutator subgroup C_4 and C_6 of the groups G_4 and G_6, then $C_4^{ab} \cong \mathbb{Z}/4\mathbb{Z}$ and $C_6^{ab} \cong \mathbb{Z}/2\mathbb{Z}$. Therefore, G_4 can not be isomorphic to G_6.

We use the computer program MAGMA to search in the maximal p-quotients of maximal class k for $1 \leq k \leq 10$ of certain finite index subgroups of the groups G_i for unmixed ramification structures.

Special Presentation G_1

There is a subgroup H_1 of index 4 in G_1 generated by b. Thus, as $H_1 \cong \mathbb{Z}$ all maximal p-quotients of p-class k of H_1 are cyclic groups of order p^k. Therefore, there will be no unmixed ramification structures coming from the groups $(H_1)_{p,k}$.

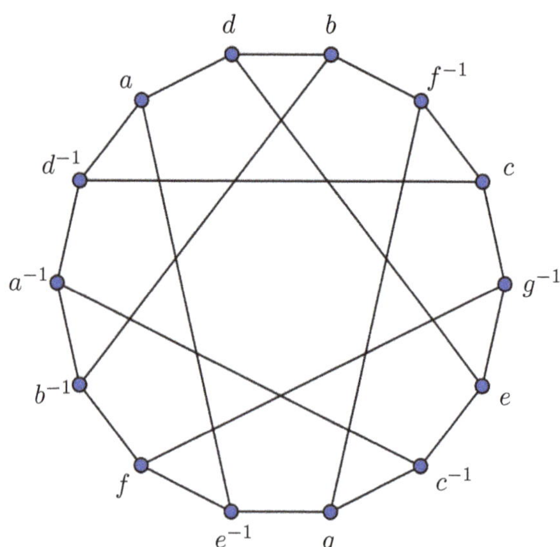

Fig. 1 Star graph of $G_1 = \langle \mathbf{x} \mid \mathbf{r}_1 \rangle$

Special Presentation G_2 and G_4

There is a subgroup H_2 of index 16 in G_2 (the commutator subgroup of G_2) generated by $h_{2,0} = bd^{-1}a^{-1}bc^{-1}$, $h_{2,1} = abd^{-1}abd^{-1}bc^{-1}$, $h_{2,2} = (bc^{-1})^2$, $h_{2,3} = adc^{-1}$ and $h_{2,4} = db^{-1}a^{-1}\,db^{-1}a^{-1}bc^{-1}$. The maximal 7-quotients of 7-class k, written as $(H_2)_{7,k}$, gives rise to a disjoint pair of spherical generators of length 3 given by the tuples,

$$T_{2,1} = [h_{2,0}, h_{2,1}, h_{2,1}^{-1}h_{2,0}^{-1}] \text{ and } T_{2,2} = [h_{2,0}h_{2,1}^2, h_{2,0}h_{2,1}^3, h_{2,1}^{-3}h_{2,0}^{-1}h_{2,2}^{-2}h_{2,0}^{-1}],$$

for $1 \leq k \leq 10$. To simplify notation, we denoted the images of $h_{2,0}$ and $h_{2,1}$ in $(H_2)_{7,k}$, again, by $h_{2,0}$ and $h_{2,1}$. Thus, the groups $(H_2)_{7,k}$ have unmixed ramification structures.

The group G_4 of [11, Example 3.3] coincides with the group C.1. in [9, Sect. 5] (via the identification $a_0 = a, a_1 = f, a_2 = c, a_3 = d, a_4 = e, a_5 = g, a_6 = b$). We find a subgroup H_4 of index 48 in G_4 generated by $h_{4,0} = da^{-1}bc^{-1}$, $h_{4,1} = bc^{-1}bc^{-1}ea^{-1}bf^{-1}$, $h_{4,2} = cb^{-1}ae^{-1}cf^{-1}$, $h_{4,3} = (ea^{-1}bf^{-1})^2$, $h_{4,4} = cb^{-1}cb^{-1}ea^{-1}bf^{-1}$ and $h_{4,5} = ea^{-1}bc^{-1}ea^{-1}bc^{-1}ea^{-1}bf^{-1}$. For $1 \leq k \leq 10$, the maximal 7-quotients of 7-class k, $(H_4)_{7,k}$, gives rise to a disjoint pair of spherical generators of length 3 given by the tuples,

$$T_{4,1} = [h_{4,0}, h_{4,1}, h_{4,1}^{-1}h_{4,0}^{-1}] \text{ and } T_{4,2} = [h_{4,0}h_{4,1}^2, h_{4,0}h_{4,1}^3, h_{4,1}^{-3}h_{4,0}^{-1}h_{4,2}^{-2}h_{4,0}^{-1}].$$

The groups H_2 and H_4 have the same maximal 7-quotients of maximal 7-class k for $1 \leq k \leq 10$. However, the abelianizations of the infinite groups are $H_2^{ab} \cong \mathbb{Z}_7 \times \mathbb{Z}_{21}$ and $H_4^{ab} \cong \mathbb{Z}_7 \times \mathbb{Z}_7$. The following theorem summarizes the above observations.

Theorem 3.2 *For $r = 2,4$, $k = 1,\ldots,10$, the groups $(H_r)_{7,k}$ are of order 7^a and admit unmixed ramification structures $(T_{r,1}, T_{r,2})$ of type $([7^b,7^b,7^b], [7^b,7^b,7^b])$ for*

$$a = \begin{cases} 2k & \text{if } k = 1, 4, 5, 8, 9, \\ 2k-1 & \text{if } k = 2, 3, 6, 7, 10. \end{cases} \quad \text{and} \quad b = \begin{cases} 2 & \text{if } 1 \leq k \leq 4, \\ 3 & \text{if } 5 \leq k \leq 8, \\ 5 & \text{if } 9 \leq k \leq 10. \end{cases}$$

This result is strong evidence that the following conjecture is true.

Conjecture 3.3 *For $r = 2,4$ and all $k \in \mathbb{N}$, the maximal 7-quotients $(H_r)_{7,k}$ of 7-class k admit unmixed ramification structures with disjoint spherical systems $(T_{r,1}, T_{r,2})$ introduced above.*

The unmixed ramification structures given for the groups $(H_2)_{7,k}$, $(H_4)_{7,k}$ above give rise to unmixed Beauville surfaces $S = (C_{T_1} \times C_{T_2})/(H_n)_{7,k}$ for $n = 2, 4$. For example, the order of the group $(H_2)_{7,1}$ and $(H_4)_{7,1}$ is 7^2. Therefore, the genera of the curves C_{T_i} is (see (3))

$$g(C_{T_1}) = g(C_{T_2}) = 1 + 2 \times 7 = 15,$$

and the holomorphic Euler-Poincaré characteristic of S is

$$\chi(S) = \frac{(g(C_{T_1}) - 1)(g(C_{T_2}) - 1)}{|G|} = 4.$$

Remark 3.4 (see [10, Beauville's Example 3.22]) The groups $(H_2)_{7,1}$, $(H_4)_{7,1}$ are isomorphic to the group $(\mathbb{Z}/7\mathbb{Z})^2$ and the two curves $C_{T_1} = C_{T_2}$ are given by the Fermat curve $x^7 + y^7 + z^7 = 0$ of degree 7. The group $(\mathbb{Z}/7\mathbb{Z})^2$ acts on $C_{T_1} \times C_{T_2}$ by the following rule

$$(\alpha, \beta) \cdot ([x : y : z], [u : v : w]) = ([\xi^\alpha x : \xi^\beta y : z], [\xi^{\alpha+2\beta} u : \xi^{\alpha+3\beta} v : w]),$$

where $\xi = e^{\frac{2\pi i}{7}}$ and $\alpha, \beta \in \mathbb{Z}/7\mathbb{Z}$. We identify $h_{n,0} \mapsto \alpha$ and $h_{n,1} \mapsto \beta$ for $n = 2, 4$.

Special Presentation G_5

The group G_5 coincides with the group A.2 in [9, Sect. 5]. We find a subgroup H_5 of index 3, generated by $h_{5,0} = ba^{-1}$, $h_{5,1} = ca^{-1}$, $h_{5,2} = da^{-1}$, $h_{5,3} = ea^{-1}$, $h_{5,4} = fa^{-1}$ and $h_{5,5} = ga^{-1}$ which have the same maximal 2-quotients of 2-class k as the group G_0 in (1) for $1 \leq k \leq 10$. However, the abelianization of this group is $H_5^{ab} \cong \mathbb{Z}_2 \times \mathbb{Z}_2 \times \mathbb{Z}_{14}$ which is not isomorphic $G_0^{ab} \cong \mathbb{Z}_2 \times \mathbb{Z}_2 \times \mathbb{Z}_6$.

In addition, we have a subgroup F_5 in H_5 of index 2, which appears to have the same maximal 2-quotients of 2-class k as H_0 (the subgroup of G_0 generated by x_0, x_1) for $1 \leq k \leq 10$. The abelianization of this group is $F_5^{ab} \cong \mathbb{Z}_4 \times \mathbb{Z}_{28}$ which is not isomorphic $G_0^{ab} \cong \mathbb{Z}_4 \times \mathbb{Z}_{14}$.

These results lead naturally to the following conjecture.

Conjecture 3.5 *Let H_5 be the index 3 subgroup of G_5 introduced above and let G_0 be the group given in (1). Let F_5 be the above mentioned index 2 subgroup of H_5, and H_0 the index 2 subgroup of G_0 generated by x_0, x_1. Even though H_5 and G_0 are not isomorphic, all corresponding maximal 2-quotients of H_5 and G_0 agree. Moreover, the same curious fact holds true for their subgroups F_5 and H_0.*

Special Presentation G_6 and G_7

The group G_6 coincides with the group B.2 in [9, Sect. 5]. We have the group specified by relations $\mathbf{r}_6 = \{a^2b, acb, bef, bge, d^2f, c^2g, egf\}$ on 7 generators but can be rewritten to a group generated by $\{x = e, y = f\}$ with relations

$$\mathbf{r}_6' = \{y^{-1}x^{-1}y^2x^{-2}y^{-3}x^{-1}, x^3yxyx^{-2}y^2\}$$

(see [3, Sect. 2.7]). The group G_7 coincides with the group B.1 in [9, Sect. 5].

We see that both groups have a subgroup H_6, H_7 of index 24 in G_6, G_7, respectively, which gives rise to maximal 3-quotients of 3-class k for $1 \le k \le 10$. However, the 3-groups are too large to successfully search for unmixed ramification structures. The abelianization of both groups is $H_6^{ab} \cong H_7^{ab} \cong \mathbb{Z}_3 \times \mathbb{Z}_3 \times \mathbb{Z}_3 \times \mathbb{Z}_3$.

3.3 Ramification Structures for the Group in [15, Example 6.4]

The group G in (2) with 13 generators $x_0, \ldots, x_{12}$ appeared as Example 6.4 in [15]. (G is constructed using Theorem 3.1 with $q = 3$, $m = 13$ and $l = \{0, 1, 3, 9\}$.) The subgroup H generated by x_0, x_1, x_2 has index 3. Again, the group G can also be found in [9, Sect. 4] as the group 1.1 (via the identification $a_i = x_{2i}$, where the indices are taken modulo 13).

The existence of ramification structures for $k = 2, \ldots, 60$ for the finite groups $H_{3,k}$ was already formulated in the Introduction (see Theorem 1.1) and was obtained by using MAGMA. See Conjecture 1.2 in the Introduction for the associated conjecture.

The ramification structures of $H_{3,k}$ in Theorem 1.1 give rise to algebraic surfaces $S = (C_{T_1} \times C_{T_2})/H_{3,k}$. For example, the order of the group $H_{3,2}$ is $a_2 = 3^6$. Therefore, the genera of the curves C_{T_i} is (see (3))

$$g(C_{T_1}) = g(C_{T_2}) = 1 + 3^5 = 244,$$

and the holomorphic Euler-Poincaré characteristic of S is

$$\chi(S) = \frac{(g(C_{T_1}) - 1)(g(C_{T_2}) - 1)}{|G|} = 81.$$

3.4 Ramification Structures for the Groups of Theorem 3.1 with $q \ge 4$

The construction given by Theorem 3.1 is for any q a prime power. For $q = 4$ the group below is given.

Example 3.6 ([15, Example 6.5]) We have that $q^2 + q + 1 = 21$ and so $\mathbb{F}_{q^3}^\times / \mathbb{F}_q^\times$ is identified with $\mathbb{Z}_{21}$. The group $\widehat{G}$ is given by the presentation,

$$\widehat{G} := \langle x_0, \ldots, x_{20} \mid x_i x_{i+7} x_{i+14}, \, x_i x_{i+14} x_{i+7}, \, x_i x_{i+3} x_{i+15} \text{ for } i \in \mathbb{Z}_{21} \rangle. \tag{4}$$

The abelianization of this group is $\widehat{G}^{ab} \cong \mathbb{Z}_2 \times \mathbb{Z}_2 \times \mathbb{Z}_2 \times \mathbb{Z}_2 \times \mathbb{Z}_6 \times \mathbb{Z}_6$. The group has maximal 2-quotients of 2-class k for $1 \le k \le 10$. However, it is extremely difficult to search for ramification structures of the maximal p-quotients of p-class k for $q \geq 4$. The finite groups are too large and have too many conjugacy classes, which leads to a computational expensive search.

3.5 Ramification Structures for the Group in [18, Example 2]

In [18], a new construction of groups presentations based on finite projective planes was introduced, generalizing the triangle presentations of [8, 9]. For the reader's convenience, we explain this briefly. The construction is based on the following general definition.

Definition 3.7 (*see* [18]) Let $\mathcal{P}_1, \ldots, \mathcal{P}_n$ be n disjoint finite projective planes of order q. Let P_i and L_i be the sets of points and lines respectively in $\mathcal{P}_i$. Let $P = \cup P_i$, $L = \cup L_i$, $P_i \cap P_j = \emptyset$ for $i \neq j$ and let λ be a bijection $\lambda : P \to L$.

A set $\mathcal{K}$ of k-tuples $(x_1, \ldots, x_k)$ will be called a *polyhedral presentation* over P compatible with λ if

(1) given $x_1, x_2 \in P$ then $(x_1, \ldots, x_k) \in \mathcal{K}$ for some $x_3, \ldots, x_k$ if and only if x_2 and $\lambda(x_1)$ are incident;
(2) $(x_1, \ldots, x_k) \in \mathcal{K}$ implies that $(x_2, \ldots, x_k, x_1) \in \mathcal{K}$;
(3) given $x_1, x_2 \in P$, then $(x_1, \ldots, x_k) \in \mathcal{K}$ for at most one $x_3 \in P$.

We call λ a *basic bijection*.

A polyhedral presentation $\mathcal{K}$ gives rise to a group presentation $G_{\mathcal{K}}$ in the following way: the generators of $G_{\mathcal{K}}$ are given by $\cup P_i$ and the relations are the k-tuples of $\mathcal{K}$, each written as a product.

Example 3.8 The *triangle presentations* listed in [9] can be seen as special cases of polyhedral presentations for $n = 1, k = 3$ and $q = 2, 3$.

We now discuss the case $n = 1, q = 2$. We enumerate the points of the projective plane by $1, 2, \ldots, 6$. The following array illustrates a basic projection λ:

$$
\begin{aligned}
0 &: 1\ 4\ 2 \\
1 &: 3\ 2\ 5 \\
2 &: 4\ 3\ 6 \\
3 &: 0\ 4\ 5 \\
4 &: 1\ 5\ 6 \\
5 &: 0\ 2\ 6 \\
6 &: 0\ 1\ 3.
\end{aligned}
$$

Here, every point k represents a row and is followed by the points contained in the associated line $\lambda(k)$. For example, the line $\lambda(3)$ consists of the points $0, 4, 5$.

A triangle presentation $\mathcal{T}$ for the group A.1 in [9] is given by

$$(0, 1, 3), (1, 2, 4), (2, 3, 5), (3, 4, 6), (4, 5, 0), (5, 6, 1), (6, 0, 2),$$

and all the cyclic permutations, i.e. for $(0, 1, 3) \in \mathcal{T}$ we also have $(1, 3, 0), (3, 0, 1)$ $\in \mathcal{T}$. The associated group presentation $G_{\mathcal{T}}$ agrees with the presentation of G_0 in (1).

Example 3.9 [18, Example 2] The projective plane $\mathcal{P}$ of order 4 can be partitioned by three projective planes of order two (see [7]). We denote points of the subplane $\mathcal{P}_i$ for $i = 1, 2, 3$ by numbers from $7i - 6$ to $7i$. Note that lines in $\mathcal{P}$ consist of five points, while the lines in $\mathcal{P}_i$ consist of three points. A basic projection λ for $\mathcal{P}$ is given below. Note that each subplane $\mathcal{P}_i$ has its own basic projection, denoted by λ_i, satisfying $\lambda_i(k) \subset \lambda(k)$. In the array below, the row associated to the point k lists first the three points in the associated line via the basic bijection in the subplane, followed up by the two remaining points in $\lambda(k)$.

<table>
<tr><td>4 : 5 6 7 12 18</td><td>9 : 12 13 14 1 15</td></tr>
<tr><td>7 : 1 2 5 8 21</td><td>11 : 8 9 12 3 17</td></tr>
<tr><td>2 : 3 4 5 14 16</td><td>14 : 10 11 12 2 16</td></tr>
<tr><td>5 : 1 3 6 10 19</td><td>12 : 8 10 13 4 18</td></tr>
<tr><td>1 : 2 4 6 9 15</td><td>10 : 9 11 13 5 19</td></tr>
<tr><td>3 : 1 4 7 11 17</td><td>13 : 8 11 13 6 20</td></tr>
<tr><td>6 : 2 3 7 13 20</td><td>8 : 9 10 14 7 21</td></tr>
</table>

$$
\begin{aligned}
&18 : 19\ 20\ 21\ \ 4\ 12 \\
&21 : 15\ 16\ 19\ \ 7\ 8 \\
&16 : 17\ 18\ 19\ \ 2\ 14 \\
&19 : 15\ 17\ 20\ \ 5\ 10 \\
&15 : 16\ 18\ 20\ \ 1\ 9 \\
&17 : 15\ 18\ 21\ \ 3\ 11 \\
&20 : 16\ 17\ 21\ \ 6\ 13
\end{aligned}
$$

The above basic projections give rise to the following polyhedral presentation $\mathcal{K}$ for a projective plane of order 4, induced by polyhedral presentations of projective planes of order 2

$$
\begin{aligned}
&(1, 9, 15), (1, 15, 9), (2, 14, 16), (2, 16, 14), (3, 11, 17), (3, 17, 11), (4, 12, 18), \\
&(4, 18, 12), (5, 10, 19), (5, 19, 10), (6, 13, 20), (6, 20, 13), (7, 8, 21), (7, 21, 8), \\
&(1, 2, 3), (1, 4, 5), (1, 6, 7), (3, 4, 6), (3, 7, 5), (2, 5, 6), (2, 4, 7), (8, 9, 12), (8, 10, 13), \\
&(8, 14, 11), (9, 14, 10), (9, 13, 11), (12, 13, 14), (10, 11, 12), (15, 16, 17), (15, 18, 19), \\
&(17, 18, 20), (17, 21, 19), (16, 19, 20), (16, 18, 21),
\end{aligned}
$$

and all their cyclic permutations.

All relators in the group presentation $G_{\mathcal{K}}$ given by $\mathcal{K}$ are of length 3 and the star graph is isomorphic to the incidence graph of a finite projective plane of order 4. This means, by [15], that the group given by this presentation $G_{\mathcal{K}}$ is a *special presentation*.

It also can be seen that this group acts on a Euclidean building where the vertex links are incidence graphs of projective planes of order 4, see [18].

There are remarkable connections between the group $G_\mathcal{K}$ and the group G_0 discussed in Sect. 3.1. Firstly, we can present the group $G_\mathcal{K}$ in an alternative way with different generators:

$$G_\mathcal{K} = \langle w_0, \ldots, w_6, y_0, \ldots, y_6, z_0, \ldots, z_6 \mid w_i w_{i+1} w_{i+3}, y_i y_{i+1} y_{i+3}, z_i z_{i+1} z_{i+3},$$
$$w_i^{-1} y_{6(1+i)} z_i^{-1}, w_i^{-1} z_i^{-1} y_{6(1+i)} (i \in \mathbb{Z}_7)\rangle, \tag{5}$$

where each of the three subsets of generators has very similar relators like those appearing for the group G_0 in (1), with only two more series of relators added representing connections between the generators of different subsets.

Secondly, the maximal 2-quotients of 2-class k of the group $G_\mathcal{K}$ are isomorphic to the maximal 2-quotients of 2-class k for the group G_0 (given by the presentation in (1)) for $1 \leq k \leq 20$. However, the groups $G_\mathcal{K}$ and G_0 are not isomorphic, as they have different abelianized groups $G_\mathcal{K}^{ab} \cong \mathbb{Z}_2 \times \mathbb{Z}_6 \times \mathbb{Z}_6$, while $G_0^{ab} \cong \mathbb{Z}_2 \times \mathbb{Z}_2 \times \mathbb{Z}_6$. This gives rise to the following conjecture.

Conjecture 3.10 *Let $G_\mathcal{K}$ be the group introduced above (e.g., by the presentation (5)), and G_0 be the group given in (1). Even though $G_\mathcal{K}$ and G_0 are not isomorphic, all corresponding maximal 2-quotients of these two groups agree.*

Remark 3.11 If we replace the relators $w_i^{-1} y_{6(1+i)} z_i^{-1}, w_i^{-1} z_i^{-1} y_{6(1+i)}$ in (5) by the relators $x_i y_i z_i, x_i z_i y_i$ we obtain a group G' with the following presentation

$$G' = \langle w_0, \ldots, w_6, y_0, \ldots, y_6, z_0, \ldots, z_6 \mid w_i w_{i+1} w_{i+3}, y_i y_{i+1} y_{i+3}, z_i z_{i+1} z_{i+3},$$
$$w_i y_i z_i, w_i z_i y_i (i \in \mathbb{Z}_7)\rangle.$$

This group G' is isomorphic to the group $\widehat{G}$ given by the presentation (4) under the identifications

$$
\begin{array}{lll}
x_0 \mapsto z_0^{-1}, & x_7 \mapsto w_0^{-1}, & x_{14} \mapsto y_0^{-1}, \\
x_1 \mapsto w_1^{-1}, & x_8 \mapsto y_1^{-1}, & x_{15} \mapsto z_1^{-1}, \\
x_2 \mapsto y_2^{-1}, & x_9 \mapsto z_2^{-1}, & x_{16} \mapsto w_2^{-1}, \\
x_3 \mapsto z_3^{-1}, & x_{10} \mapsto w_3^{-1}, & x_{17} \mapsto y_3^{-1}, \\
x_4 \mapsto w_4^{-1}, & x_{11} \mapsto y_4^{-1}, & x_{18} \mapsto z_4^{-1}, \\
x_5 \mapsto y_5^{-1}, & x_{12} \mapsto z_5^{-1}, & x_{19} \mapsto w_5^{-1}, \\
x_6 \mapsto z_6^{-1}, & x_{13} \mapsto w_6^{-1}, & x_{20} \mapsto y_6^{-1}.
\end{array}
$$

Acknowledgments We thank Donald Cartwright for the representations and method given in Appendix 2 and helpful correspondences. The first author also wishes to thank Uzi Vishne for useful correspondences.

Appendix 1: Expanders Associated to the Group G_0

Expander graphs are defined with the help of the edge expansion rate.

Definition A. 1 Let $\mathcal{G} = (V, E)$ be a combinatorial graph with vertex set V and edge set E. Then the *edge expansion rate* $h(\mathcal{G})$ is defined as

$$h(\mathcal{G}) = \inf_{\text{finite} A \subset V} \frac{|\partial A|}{\min(|A|, |V \setminus A|)},$$

where $\partial A \subset E$ is the set of all edges connecting a vertex of A with a vertex of $V \setminus A$.

Expanders are infinite families of finite graphs which are both sparse and highly connected. They are not only theoretically important but have also applications in computer science for, e.g., robust network designs.

Definition A. 2 A sequence $\mathcal{G}_n = (V_n, E_n)$ of connected finite graphs with $|V_n| \to \infty$ is called a *family of expanders* if there exists $k \geq 2$ and $\epsilon > 0$ such that

(a) all graphs $\mathcal{G}_n$ are k-regular,
(b) $h(\mathcal{G}_n) \geq \epsilon$ for all n.

It was observed in [22] that the subgroup H_0 of G_0 generated by x_0, x_1 has index 2, and that both groups H_0 and G_0 are just infinite and have Kazhdan property (T). Property (T) implies that, for a fixed choice of generators, the Cayley graphs of all quotients by finite index normal subgroups have a uniform positive lower bound for their edge expansion rate (see [19, Proposition 3.3.1]). A presentation of the subgroup H_0 is given by

$$H_0 = \langle x_0, x_1 \mid r_1, r_2, r_3 \rangle,$$

where

$$r_1 = (x_1 x_0)^3 x_1^{-3} x_0^{-3},$$
$$r_2 = x_1 x_0^{-1} x_1^{-1} x_0^{-3} x_1^2 x_0^{-1} x_1 x_0 x_1,$$
$$r_3 = x_1^3 x_0^{-1} x_1 x_0 x_1 x_0^2 x_1^2 x_0 x_1 x_0.$$

We have the following Cayley graph expanders obtained from finite groups with just two generators and four relations.

Theorem A. 3 (cf. [22, Theorem 1]) *The groups*

$$H_k = \langle x_0, x_1 \mid r_1, r_2, r_3, [x_1, \underbrace{x_0, \ldots, x_0}_{k}] \rangle$$

are finite with $|H_k| \to \infty$, and the associated Cayley graphs with respect to the generators x_0, x_1 define an infinite family of expanders of vertex degree 4.

Using the faithful matrix representation of H_0 by infinite upper triangular matrices and their truncations at the kth upper diagonal as mentioned in Sect. 3.1, we obtain another family $\widetilde{H}_k$ of finite *nilpotent* groups whose associated Cayley graphs $\mathcal{G}_k$ with respect to the generators x_0, x_1 are another family of expander graphs which form a tower of coverings

$$\cdots \mathcal{G}_k \to \mathcal{G}_{k-1} \to \cdots \to \mathcal{G}_1 \to \mathcal{G}_0,$$

whose covering indices are powers of 2 (for more details, see [22]). It was conjectured in [22, Conjecture 2] that the covering indices follow the pattern 4, 8, 4, 8, 8, 4, 8, 8, 4, 8, 8, ... See Fig. 2 for the graph $\mathcal{G}_2$. We use the notation $z_1 = [x_0, x_1]$, $z_2 = x_0^2$, $z_3 = x_1^2$ and $z_{ij} = z_i z_j$ and $z_{ijk} = z_i z_j z_k$. The elements expressed by z_i lie in the centre of $\widetilde{H}_2$. The same graph was illustrated in [22, Fig. 4], but the illustration given here is more symmetric. Solid edges from vertices with label i to vertices with label $i + 1 \pmod 4$ represent right multiplication by x_0, while dashed edges from vertices with label i to vertices with label $i + 1 \pmod 4$ represent right multiplication by x_1. Note that the solid 4-cycles as well as the dashed 4-cycles in $\mathcal{G}_3$ are consequences of $x_0^4 = x_1^4 = 1$ in $\widetilde{H}_2$.

Another construction of 3-regular expanders was given in [17]. Starting from the same groups $\widetilde{H}_k$, we now consider the associated Cayley graphs X_k with respect to

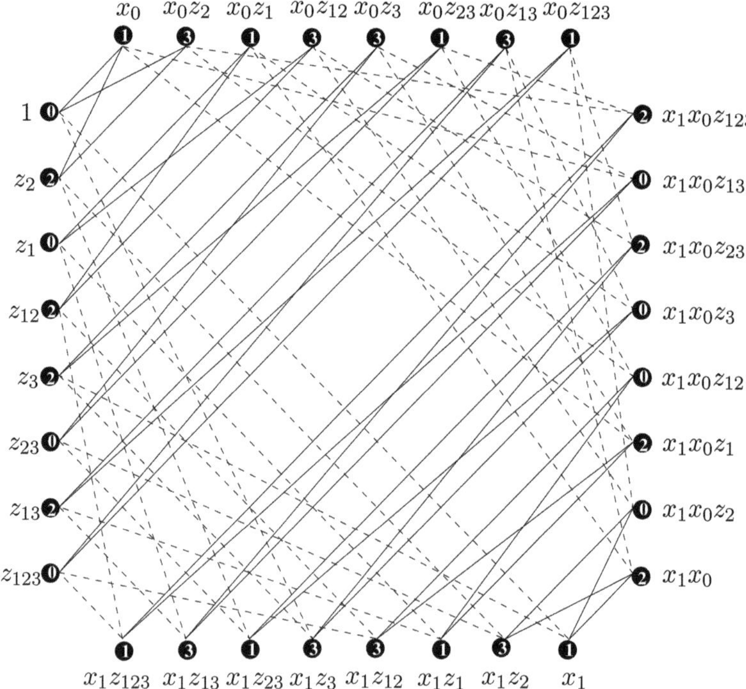

Fig. 2 The graph $\mathcal{G}_2$

the generators x_0, x_1, x_3 where $x_3 = x_1^{-1} x_0^{-1}$. The graphs X_k are 6-regular and $\mathcal{G}_k$ is a subgraph of X_k with the same number of vertices. Property (T) guarantees that the graphs X_k are also a family of expanders. One can check that X_k forms a tessellation of a closed Riemann surface by triangles and by 2^l-gons (with l only depending on k) and that every edge of X_k belongs to precisely one triangle of X_k. Now we apply a $\Delta - Y$ transformation to the graphs X_k to obtain new graphs T_k. The $\Delta - Y$ transformation removes the edges of every triangle in the original graph X_k, adds a new vertex in its centre, and connects this new central vertex with new edges to the 3 original vertices of the triangle. It turns out that the vertex set of the new graph T_k is twice as large as the vertex set of old graph X_k, and that T_k is 3-regular. Moreover, there is an explicit connection between the eigenfunctions of the adjacency matrix of X_k and the eigenfunctions of the adjacency matrix of T_k (see [17, Theorem 2.1]). The spectral characterisation of expander graphs then implies that the new family T_k of 3-regular graphs is, again, a family of expanders.

For yet another expander graphs construction from the group G_0 see [20, 23].

Appendix 2: Representation for the Group G

We include a representation for the group G (given by (2)) in $GL(9, \mathbb{F}_3[1/Y])$, which may be useful in the future (as the matrix representations for the group G_0 with presentation (1) were useful for several works [8, 20, 22]). The representation is due to Donald Cartwright and the algebra program REDUCE. Recall that the group G coincides with the group 1.1 in [9], where we relate the generators by $a_i = x_{2i}$ for $i = 0, \ldots, 12$, with indices taken modulo 13. We set

$$
x_0 : \begin{pmatrix}
1 & 1 & 1 & 0 & 2 & 2 & 0 & 1 & 1 \\
0 & 1 & 2 & 0 & 0 & 1 & 0 & 0 & 2 \\
0 & 0 & 1 & 0 & 0 & 0 & 0 & 0 & 0 \\
0 & 0 & 0 & 1 & 1 & 1 & 0 & 2 & 2 \\
0 & 0 & 0 & 0 & 1 & 2 & 0 & 0 & 1 \\
0 & 0 & 0 & 0 & 0 & 1 & 0 & 0 & 0 \\
0 & 0 & 0 & 0 & 0 & 0 & 1 & 1 & 1 \\
0 & 0 & 0 & 0 & 0 & 0 & 0 & 1 & 2 \\
0 & 0 & 0 & 0 & 0 & 0 & 0 & 0 & 1
\end{pmatrix}
+ \frac{1}{Y}
\begin{pmatrix}
0 & 2 & 2 & 0 & 1 & 1 & 0 & 2 & 2 \\
0 & 0 & 1 & 0 & 0 & 2 & 0 & 0 & 1 \\
0 & 0 & 0 & 0 & 0 & 0 & 0 & 0 & 0 \\
0 & 1 & 1 & 0 & 2 & 2 & 0 & 1 & 1 \\
0 & 0 & 2 & 0 & 0 & 1 & 0 & 0 & 2 \\
0 & 0 & 0 & 0 & 0 & 0 & 0 & 0 & 0 \\
0 & 2 & 2 & 0 & 1 & 1 & 0 & 2 & 2 \\
0 & 0 & 1 & 0 & 0 & 2 & 0 & 0 & 1 \\
0 & 0 & 0 & 0 & 0 & 0 & 0 & 0 & 0
\end{pmatrix}
$$

and

$$\tau : \begin{pmatrix} 1 & 0 & 0 & 0 & 0 & 0 & 0 & 0 & 0 \\ 0 & 1 & 0 & 0 & 0 & 0 & 0 & 0 & 0 \\ 0 & 0 & 1 & 0 & 0 & 0 & 0 & 0 & 0 \\ 0 & 0 & 0 & 0 & 1 & 0 & 0 & 0 & 0 \\ 0 & 0 & 0 & 0 & 1 & 1 & 0 & 0 & 0 \\ 0 & 0 & 0 & 1 & 0 & 1 & 0 & 0 & 0 \\ 0 & 0 & 0 & 0 & 0 & 0 & 2 & 2 & 0 \\ 0 & 0 & 0 & 0 & 0 & 0 & 0 & 1 & 2 \\ 0 & 0 & 0 & 0 & 0 & 0 & 2 & 0 & 1 \end{pmatrix},$$

where the other generators $x_1, \ldots, x_{12}$ are formed via conjugation of x_0 by τ, i.e. $x_i = \tau^i x_0 \tau^{-i}$ for $i = 1, \ldots, 12$.

The idea in creating this representation is to write $\mathbb{F}_{27} = \mathbb{F}_3(\theta)$, where θ is a primitive element on $\mathbb{F}_{27}$ satisfying $\theta^3 = \theta + 1$, and to use the basis $\{\theta^i \sigma^j | i, j = 0, 1, 2\}$ for the divison algebra $\mathcal{A}$ over $\mathbb{F}_{27}(Y)$ for an indeterminate Y (in the order $1, \theta, \theta^2, \sigma, \theta\sigma, \ldots, \theta^2\sigma^2$). Here σ is assumed to satisfy $\sigma^3 = Y - 1$ (which implies $(1 + \sigma)^{-1} = (1/Y)(1 - \sigma + \sigma^2)$) and $\sigma\theta\sigma^{-1} = \theta^3$. The generators of $\mathcal{T}_{\mathcal{K}}$, where $\mathcal{K}$ is a triangle presentation from [8, 9], are the $a_u = u^{-1}(1 + \sigma)u$, where $u \in \mathbb{F}_{27}^\times / \mathbb{F}_3^\times$. Since $\mathbb{F}_{27}^\times = \mathbb{F}_3^\times \cdot \{1 = \theta^{13}, \theta, \ldots, \theta^{12}\}$, we choose $\alpha_k = \theta^{-k}(1 + \sigma)\theta^k$ as in [9, p. 178]. The α_k's act on $\mathcal{A}$ by conjugation. A straightforward calculation yields

$$\alpha_k \theta^i \sigma^j \alpha_k^{-1} = \theta^i \sigma^j \frac{1}{Y} + \left(\theta^{3i+2k} - \theta^{i+2\cdot3^j k} \right) \sigma^{j+1} \frac{1}{Y}$$
$$+ \left(\theta^{i+8\cdot3^j k} - \theta^{3i+2k+2\cdot3^{j+1}k} \right) \sigma^{j+2} \frac{1}{Y} + \theta^{3i+2k+8\cdot3^{j+1}k} \sigma^j \frac{Y-1}{Y}.$$

Expressing the conjugation by α_k with respect to the above basis of $\mathcal{A}$ then gives rise to a representation as a 9×9 matrix over the field $\mathbb{F}_3(1/Y)$. We conclude from [9] that the matrices associated to the α_k satisfy the relations of our generators x_k. Note, finally, that the above matrix for τ represents the conjugation by θ in $\mathcal{A}$, i.e., $z \mapsto \theta^{-1} z \theta$.

References

1. N. Barker, N. Boston, N. Peyerimhoff, A. Vdovina, New examples of Beauville surfaces. Monatsh. Math. **166**(3–4), 319–327 (2012)
2. N. Barker, N. Boston, N. Peyerimhoff A. Vdovina, An infinite family of 2-groups with mixed Beauville structures. Int. Math. Res. Not. (2014). doi:10.1093/imrn/rnu045
3. S. Barre, Polyédres de rang deux, Thesis ENS Lyon, December (1996). http://web.univ-ubs.fr/lmam/barre/these1.pdf
4. I.C. Bauer, F. Catanese F. Grunewald, Beauville surfaces without real structures, in *Geometric Methods in Algebra and Number Theory*. Progress in Mathematics, vol. 235 (Birkhäuser, Boston, 2005)
5. I.C. Bauer, F. Catanese, F. Grunewald, The classification of surfaces with $p_g = q = 0$ isogenous to a product of curves. Pure Appl. Math. Q. **4**(2), 547–586 (2008)

6. W. Bosma, J. Cannon, C. Playoust, The Magma algebra system. I. The user language. J. Symb. Comput. **24**(3–4), 235–265 (1997)
7. R.H. Bruck, Quadratic extensions of cyclic planes, in *Proceedings Symposium in Applied Mathematics*, pp. 15–44 (1960)
8. D.I. Cartwright, A.M. Mantero, T. Steger, A. Zappa, Groups acting simply transitively on the vertices of a building of type $\widetilde{A}_2$, I. Geom. Dedicata **47**(2), 143–166 (1993)
9. D.I. Cartwright, A.M. Mantero, T. Steger, A. Zappa, Groups acting simply transitively on the vertices of a building of type $\widetilde{A}_2$, II. Geom. Dedicata **47**(2), 167–223 (1993)
10. F. Catanese, Fibred surfaces, varieties isogenous to a product and related moduli spaces. Am. J. Math. **122**(1), 1–44 (2000)
11. M. Edjvet, J. Howie, Star graphs, projective planes and free subgroups in small cancellation groups. Proc. Lond. Math. Soc. **57**(2), 301–328 (1988)
12. M. Edjvet, A. Vdovina, On the SQ-universality of groups with special presentations. J. Group Theory **13**(6), 923–931 (2010)
13. B. Fairbairn, Some exceptional Beauville structures. J. Group Theory **15**, 631–639 (2012)
14. Y. Fuertes, G. González-Diez, On Beauville structures on the groups S_n and A_n. Math. Z. **264**, 959–968 (2010)
15. J. Howie, On the SQ-universality of $T(6)$-groups. Forum Math. **1**(3), 251–272 (1989)
16. D.R. Hughes, F.C. Piper, *Projective Planes* (Springer, New York, 1973)
17. I. Ivrissimtzis, N. Peyerimhoff, A. Vdovina, Trivalent expandes and hyperbolic surfaces (2012) arXiv:1202.2304
18. S. Immervol, A. Vdovina, Partitions of projective planes and construction of polyhedra, Max-Planck-Institut fur Mathematik, Bonn, Preprint Series 23 (2001)
19. A. Lubotzky, *Discrete Groups, Expanding Graphs and Invariant Measures* (Birkhäuser, Basel, 2010)
20. A. Lubotzky, B. Samuels, U. Vishne, Explicit construction of Ramanujan complexes of type $\widetilde{A}_d$. Eur. J. Comb. **26**(6), 965–993 (2005)
21. R.C. Lyndon, P.E. Schupp, *Combinatorial Group Theory, Classics in Mathematics*. Reprint of the 1977 edition (Springer, Berlin, 2001)
22. N. Peyerimhoff, A. Vdovina, Cayley graph expanders and groups of finite width. J. Pure Appl. Algebra **215**(11), 2780–2788 (2011)
23. A. Sarveniazi, Explicit construction of a Ramanujan $(n_1, n_2, \ldots, n_{d-1})$-regular hypergraph. Duke Math. J. **139**(1), 141–171 (2007)
24. F. Serrano, Isotrivial fibred surfaces. Ann. Mat. Pura Appl. **171**(4), 63–81 (1996)

A Survey of Beauville p-Groups

Nigel Boston

Abstract This paper describes recent results as to which p-groups are Beauville, with emphasis on ones of small order (joint with N. Barker and B. Fairbairn) and ones that form inverse systems (joint with N. Barker, N. Peyerimhoff, and A. Vdovina).

2000 Mathematics Subject Classification. 20D15 $\cdot$ 14J29 $\cdot$ 14L30 $\cdot$ 20F05

1 Introduction

Much research has gone into the fundamental question of which finite groups are Beauville. This has been carried out mostly for simple (and related) groups. In a sense, however, most finite groups are p-groups (for example, about 99.2 % of the groups of order ≤ 2000 have order 2^{10}). Moreover, as is described below, p-groups form a class of groups with a great deal of structure. It is, therefore, natural to consider which finite p-groups are Beauville.

Until very recently, this would have made for a very short survey. The seminal work of Bauer, Catanese, and Grunewald [5, 6] included the classification of all abelian Beauville groups. Among these, the p-groups arising are precisely $\mathbf{Z}/p^k \times \mathbf{Z}/p^k$ for $p \geq 5$ and any $k \geq 1$. There have, however, been only a handful of nonabelian Beauville p-groups documented, in [6], where two examples of order 2^8 were presented, and by Fuertes, González-Diez, and Jaikin-Zapirain [8], who gave examples of order 2^{12} and 3^{12}. It turns out, however, that there is much more to be said.

After recalling the definition of a Beauville group, this paper begins with a description of the now standard way to organize the class of finite 2-generator p-groups into a rooted tree. This then facilitates (computational and theoretical) investigations into the location of Beauville groups within this tree. We focus on three particular aspects, namely whether being Beauville or not is inherited by neighbours in the tree, the

N. Boston (✉)
Department of Mathematics, University of Wisconsin, 303 Van Vleck Hall, 480 Lincoln Drive, Madison, WI 53706, USA
e-mail: boston@math.wisc.edu

© Springer International Publishing Switzerland 2015

I. Bauer et al. (eds.), *Beauville Surfaces and Groups*, Springer Proceedings in Mathematics & Statistics 123, DOI 10.1007/978-3-319-13862-6_3

existence of infinite paths through the tree consisting entirely or mostly of Beauville or non-Beauville groups, and the density of Beauville groups among groups of order p^n for fixed n as $p \to \infty$.

2 Definitions

2.1 Beauville Structures

Definition 2.1 (*Unmixed Beauville structure*) Let G be a finite group. Given $x, y \in G$, let $\Sigma(x, y)$ denote the union of all conjugacy classes of all powers of x, y, xy. In the unmixed case, we call G *Beauville* if it has two pairs of generators, $\{x, y\}$ and $\{u, v\}$, such that the following holds:

$$(*) \quad \Sigma(x, y) \cap \Sigma(u, v) = \{1\}$$

In terms of (unmixed) Beauville surfaces minimally represented as $(C_1 \times C_2)/G$ where the map $C_i \to C_i/G = \mathbf{P}^1(\mathbf{C})$ is ramified over $0, 1, \infty$ ($i = 1, 2$), the fact that G is 2-generated is equivalent to it being a quotient of $\pi_1(\mathbf{P}^1(\mathbf{C}) - \{0, 1, \infty\})$ whilst $(*)$ is equivalent to G acting freely on $C_1 \times C_2$.

In the mixed case, the group G can interchange the curves C_i and then the subgroup H of transformations stabilizing each curve has index 2 in G.

Definition 2.2 (*Mixed Beauville structure*) We call G (mixed) *Beauville* if it has a subgroup H of index 2 which has a pair of generators $T = \{x, y\}$ such that, for every $g \in G - H$, $\Sigma(T) \cap \Sigma(gTg^{-1}) = \{1\}$ and $g^2 \notin \Sigma(T)$.

Note that if G is a p-group with mixed Beauville structure, then necessarily $p = 2$.

2.2 p-Groups

As for the basics on p-groups, we focus on a certain filtration by normal subgroups.

Definition 2.3 (*Lower p-central series*) If G is a finite p-group, then it has a normal series

$$\{1\} = P_c(G) \subseteq \dots \subseteq P_1(G) \subseteq P_0(G) = G,$$

called the *lower p-central series* of G, given by $P_{i+1}(G) = [G, P_i(G)]P_i(G)^p$. The smallest c such that $P_c(G) = \{1\}$ is called the *p-class* of G.

The first term $P_1(G)$ is sometimes called the *Frattini subgroup* of G. It is best to think in terms of the corresponding sequence of quotients of G,

$$G = G/P_c(G) \to G/P_{c-1}(G) \to \ldots \to G/P_1(G) = (\mathbf{Z}/p)^{d(G)}$$

where $d(G)$ denotes the *minimum number of generators* of G. In this paper we will be exclusively interested in the case $d(G) = 2$. For each p we will create a rooted tree in which the arrows above correspond to edges leading to the root $(\mathbf{Z}/p)^2$. The group $G/P_1(G)$ is elementary p-abelian by the definition of $P_1(G)$—its rank is $d(G)$ by the following classical result.

Lemma 2.4 (Burnside's Basis Theorem) $x, y \in G$ *generate* G *if and only if their images in* $G/P_1(G)$ *generate* $G/P_1(G)$.

Suppose we fix p and d. Then there is a rooted tree, due to O'Brien [11], whose vertices are the isomorphism classes of finite p-groups G with $d(G) = d$. A group G of p-class c is joined to all groups H of p-class $c + 1$ with $H/P_c(H) \cong G$ (these are called the *children* of G). If G has minimal presentation F/R, then these are constructed as quotients of a covering group $G^* = F/R^*$. To each finite p-group is attached a nonnegative integer called its *nuclear rank*, which is defined to be the rank of the elementary abelian group $P_c(G^*)$. This is 0 if and only if the group has no children. See [11] for the technical details.

3 Beauville Groups in O'Brien Trees

3.1 An Example

To illustrate the general theory, consider the case of $p = 5, d = 2$. The root of the O'Brien tree is $\mathbf{Z}/5 \times \mathbf{Z}/5$, which is Beauville as noted in the introduction (indeed it is the first example, found by Beauville himself). It has seven children, of which three have order 5^3, three order 5^4, and one order 5^5 (namely the free 2-generator 5-group of 5-class 2). Of these, we compute using the computer algebra package MAGMA [7] that there are four Beauville groups, namely one of order 5^3, two of order 5^4 (including $\mathbf{Z}/25 \times \mathbf{Z}/25$), and the one of order 5^5.

Consider the three groups of order 5^3. One is $\mathbf{Z}/5 \times \mathbf{Z}/25$, which is not Beauville. It has two children, one childless, the other having two children, one childless, and so on. None of these groups is Beauville. The other group of order 5^3 that is not Beauville is childless. Let us focus on the remaining group of order 5^3, which is Beauville. It has 16 children, four of order 5^4 (one Beauville, three not) and twelve of order 5^5 (nine Beauville, three not).

Note that a child of a Beauville group may or may not be Beauville. As seen with the first group of order 5^3 above, a child of a group that is not Beauville can also not be Beauville. Since $\mathbf{Z}/3 \times \mathbf{Z}/3$ is not Beauville, yet Beauville 3-groups exist, there must exist a group that is not Beauville but which has a Beauville child. There does, however, appear to be some sort of correlation between nuclear rank and being Beauville, as I now explain.

Of the 16 children in the last but one paragraph, one has nuclear rank 3. It is Beauville of order 5^5. It has 116 children. Of these, 8 have order 5^6—the 5 with nuclear rank 0 are not Beauville, the rest are. Of the 116 groups, 61 have order 5^7—the 17 with nuclear rank 2 are not Beauville, the rest are. Of the 116 groups, 47 have order 5^8—the 5 with nuclear rank 4 are not Beauville, the rest are.

There is no exact rule—for instance, of the 16 children in the last but second paragraph, there are five of order 5^5 with nuclear rank 1, two Beauville, three not—but clearly something is going on here, possibly for large enough p-class.

We gather our conclusions here:

Theorem 3.1 *A child of a Beauville p-group may or may not be Beauville. A child of a non-Beauville p-group may or may not be Beauville. There exists a p-group all of whose infinitely many descendants are not Beauville.*

3.2 Infinite Paths Through the O'Brien Trees

The infinite ends of the O'Brien tree correspond to infinite 2-generator pro-p groups, which are obtained as the inverse limit of the groups in the path. Alternatively one might start with a discrete 2-generator group and investigate its pro-p completion.

In this way it was found that, along an infinite ray of the tree, the property of being Beauville or not can alternate infinitely often, as noted in my joint work with Barker, Peyerimhoff, and Vdovina [2, 3]. We considered the groups

$$H = \langle x_0, x_1 \rangle \subseteq \Gamma = \langle x_0, ..., x_6 \mid x_i x_{i+1} x_{i+3} = 1, i \in \mathbf{Z}/7 \rangle$$

and conjectured that if k is not a power of 2, then $H/P_k(H)$ is unmixed Beauville and $H/P_k(H) \subseteq G/P_k(G)$ is mixed Beauville. This was checked computationally for several k [2] and proven in general in [3]. Note that the case $k = 3$ gives one of the two mixed Beauville groups of order 256 originally given in [6], SmallGroup (256, 3679), the other being SmallGroup (256, 3678).

In [4], we also found similar results for other groups with special presentation (defined by Howie [10] to be groups for which every relator has length 3 and the star graph is isomorphic to the incidence graph of a finite projective plane). There are also simpler examples to be had, for instance by exploiting the lemma below of Fuertes and Jones.

Lemma 3.2 ([9, Lemma 4.2]) *If $x, y, u, v \in G$ have images $\overline{x}, \overline{y}, \overline{u}, \overline{v} \in G/N$ yielding a Beauville structure on G/N, then if*

$$\langle x \rangle \cap N = \langle y \rangle \cap N = \langle xy \rangle \cap N = \{1\},$$

then x, y, u, v yield a Beauville structure on G.

An easy induction immediately gives:

Corollary 3.3 *If $p \geq 5$ and Γ is the triangle group*

$$\langle x, y, z \mid x^p = y^p = z^p = xyz = 1 \rangle,$$

then its p-central quotients $\Gamma/P_k(\Gamma)$ are all Beauville.

I would conjecture that likewise:

Conjecture 3.4 *If $p \geq 5$ and Γ is either*

(i)   the free product $\langle x, y \mid x^p, y^p \rangle$ or
(ii) the free group $\langle x, y \rangle$,
      then its p-central quotients $\Gamma/P_k(\Gamma)$ are all Beauville.

4 Beauville Groups of Small Order

A basic question often asked is: are most 2-generated groups Beauville? Since the evidence overwhelmingly indicates that most groups (2-generated or otherwise) are p-groups, we address this question here. In [8], the authors said it was very plausibly true that most 2-generator p-groups are Beauville. One way to make sense of the question is to see whether this holds for groups of order p^n for fixed n as p varies. This is joint work with Barker and Fairbairn [1].

To begin with, we sought the smallest (nonabelian) Beauville p-group for each p.

Theorem 4.1 ([1, Corollary 1.9])

(1)   The smallest Beauville 2-group is SmallGroup(128, 36) (also discovered in [6]);
(2)   the smallest Beauville 3-group is SmallGroup(243, 3);
(3)   for $p \geq 5$ the smallest nonabelian Beauville p-group is SmallGroup(p^3, 3).

Definition 4.2 Let $g_n(p)$ denote the number of groups G of order p^n with $d(G) = 2$. Let $h_n(p)$ denotes the number of Beauville groups of order p^n. Let $\delta_n(p) = h_n(p)/g_n(p)$, which denotes the density of Beauville groups among all 2-generated groups of order p^n.

We found using the computer algebra package MAGMA [7] that:

Theorem 4.3
$$g_2(p) = 1; \quad h_2(p) = 1 \text{ if } p \geq 5$$

$$g_3(p) = 3; \quad h_3(p) = 1 \text{ if } p \geq 5$$

$$g_4(p) = 9 \text{ if } p \geq 3; \quad h_4(p) = 3 \text{ if } p \geq 5$$

$$g_5(p) = p + 26 + 2gcd(p - 1, 3) + gcd(p - 1, 4); \quad h_5(p) \geq p + 8 \; if \; p \geq 5$$

$$g_6(p) = 10p + 62 + 14gcd(p - 1, 3) + 7gcd(p - 1, 4) + 2gcd(p - 1, 5);$$

$$h_6(p) \leq g_6(p) - (p - 1) \; if \; p \geq 5$$

Corollary 4.4 *The limit as $p \rightarrow \infty$ of $\delta_n(p)$ is 1 if $n = 2$ or 5, but strictly less than 1 if $n = 3, 4,$ or 6.*

Conjecture 4.5 $h_5(p) = p + 10$ *if $p \geq 5$.*

There is clearly plenty of room for further investigations into the questions above. Preliminary calculations suggest that the limit is less than 1 if $n = 7$.

Acknowledgments I thank my co-authors Nathan Barker, Ben Fairbairn, Norbert Peyerimhoff, and Alina Vdovina for their encouragement and hard work in helping develop this field.

References

1. N. Barker, N. Boston, B. Fairbairn, A note on Beauville p-groups. Exp. Math. **21**(3), 298–306 (2012)
2. N. Barker, N. Boston, N. Peyerimhoff, A. Vdovina, New examples of Beauville surfaces. Monatsh. Math. **166**(3–4), 319–327 (2012)
3. N. Barker, N. Boston, N. Peyerimhoff, A. Vdovina, An infinite family of 2-groups with mixed Beauville structures (2013). arXiv:1304.4480
4. N. Barker, N. Boston, N. Peyerimhoff, A. Vdovina, Regular algebraic surfaces, ramification structures, and projective planes, submitted to this volume
5. I.C. Bauer, F. Catanese, F. Grunewald, Beauville surfaces without real structures, in *Geometric Methods in Algebra and Number Theory*. Progress in Mathematics, vol. 235 (Birkhäuser Boston, 2005)
6. I.C. Bauer, F. Catanese, F. Grunewald, The classification of surfaces with $p_g = q = 0$ isogenous to a product of curves. Pure Appl. Math. Q. **4**(2), 547–586 (2008)
7. W. Bosma, J. Cannon, C. Playoust, The Magma algebra system, I. The user language. J. Symb. Comput. **24**(3–4), 235–265 (1997)
8. Y. Fuertes, G. González-Diez, A. Jaikin-Zapirain, On Beauville surfaces. Groups Geom. Dyn. **5**(1), 107–119 (2011)
9. Y. Fuertes, G.A. Jones, Beauville surfaces and finite groups. J. Algebra **340**(1), 13–27 (2011)
10. J. Howie, On the SQ-universality of $T(6)$-groups. Forum Math. **1**(3), 251–272 (1989)
11. E.A. O'Brien, The p-group generation algorithm. J. Symb. Comput. **9**, 677–698 (1990)

Strongly Real Beauville Groups

Ben Fairbairn

Abstract A strongly real Beauville group is a Beauville group that defines a real Beauville surface. Here we discuss efforts to find examples of these groups, emphasising on the one extreme finite simple groups and on the other abelian and nilpotent groups. We will also discuss the case of characteristically simple groups and almost simple groups. *En route* we shall discuss several questions, open problems and conjectures as well as giving several new examples of infinite families of strongly real Beauville groups.

1 Introduction

We first issue an apology/assurance. It is the nature of Beauville constructions that this article is likely to be of interest to both geometers and group theorists. The author is painfully aware of this. As a consequence there will be times when we make statements that may seem obvious or elementary to the group theorist but may seem quite surprising to the geometer.

We begin with the usual definitions to establish notation and terminology.

Definition 1 A surface $\mathcal{S}$ is a **Beauville surface of unmixed type** if

- the surface $\mathcal{S}$ is isogenous to a higher product, that is, $\mathcal{S} \cong (\mathcal{C}_1 \times \mathcal{C}_2)/G$ where $\mathcal{C}_1$ and $\mathcal{C}_2$ are algebraic curves of genus at least 2 and G is a finite group acting faithfully on $\mathcal{C}_1$ and $\mathcal{C}_2$ by holomorphic transformations in such a way that it acts freely on the product $\mathcal{C}_1 \times \mathcal{C}_2$ and
- each $\mathcal{C}_i/G$ is isomorphic to the projective line $\mathbb{P}_1(\mathbb{C})$, and the covering map $\mathcal{C}_i \to \mathcal{C}_i/G$ is ramified over three points.

What makes these surfaces so easy to work with is the fact that the definition above can be translated into purely group theoretic terms—the following definition imposes equivalent conditions on the group G.

B. Fairbairn (✉)
Department of Economics, Mathematics and Statistics, Birkbeck,
University of London, Malet Street, London WC1E 7HX, UK
e-mail: b.fairbairn@bbk.ac.uk

© Springer International Publishing Switzerland 2015

I. Bauer et al. (eds.), *Beauville Surfaces and Groups*, Springer Proceedings
in Mathematics & Statistics 123, DOI 10.1007/978-3-319-13862-6_4

Definition 2 Let G be a finite group. Let $x, y \in G$ and let

$$\Sigma(x, y) := \bigcup_{i=1}^{|G|} \bigcup_{g \in G} \{(x^i)^g, (y^i)^g, ((xy)^i)^g\}.$$

An **unmixed Beauville structure** for G is a pair of generating sets of elements $\{\{x_1, y_1\}, \{x_2, y_2\}\} \subset G \times G$ such that $\langle x_1, y_1 \rangle = \langle x_2, y_2 \rangle = G$ and

$$\Sigma(x_1, y_1) \cap \Sigma(x_2, y_2) = \{e\} \tag{†}$$

where e denotes the identity element of G. If G has a Beauville structure, then we say that G is a **Beauville group**. Furthermore we say that the structure has **type** $((o(x_1), o(y_1), o(x_1y_1)), (o(x_2), o(y_2), o(x_2y_2)))$.

In the author's experience, upon seeing the above definition, group theorists often retort "why record just the orders of the elements and not precisely which classes the elements belong to?" Determining precisely which class an element belongs to is much harder than determining its order. Furthermore, in practice, when ensuring that a set of elements satisfies condition (†) the easiest way often is to show that $o(x_1)o(y_1)o(x_1y_1)$ is coprime to $o(x_2)o(y_2)o(x_2y_2)$. This simple observation has been used to great effect by several authors—see [18, 20, 22, 25] among others. Furthermore, the type alone encodes substantial amounts of geometric information: the Riemann-Hurwitz formula

$$g(\mathcal{C}_i) = 1 + \frac{|G|}{2}\left(1 - \frac{1}{o(x_i)} - \frac{1}{o(y_i)} - \frac{1}{o(x_iy_i)}\right)$$

tells us the genus of each of the curves used to define the surface $\mathcal{S}$. Indeed, whilst some groups have generating pairs that by the above formula define surfaces with the property that $g(\mathcal{C}) \leq 1$, condition (†) ensures that for each i we have that $g(\mathcal{C}_i) \geq 2$. Furthermore, a theorem of Zeuthen-Segre also gives us the Euler number of the surface $\mathcal{S}$ since

$$e(\mathcal{S}) = 4\frac{(g(\mathcal{C}_1) - 1)(g(\mathcal{C}_2) - 1)}{|G|},$$

which in turn gives us the holomorphic Euler-Poincaré characteristic of $\mathcal{S}$ from the relation $4\chi(\mathcal{S}) = e(\mathcal{S})$—see [14, Theorem 3.4].

In light of the above, we make the following non-standard definition which will be of use in what follows.

Definition 3 We say that a Beauville structure $\{\{x_1, y_1\}, \{x_2, y_2\}\}$ is **coprime** if $o(x_1)o(y_1)o(x_1y_1)$ and $o(x_2)o(y_2)o(x_2y_2)$ are coprime.

Given any complex surface $\mathcal{S}$ it is natural to consider the complex conjugate surface $\overline{\mathcal{S}}$. In particular, it is natural to ask whether the surfaces are biholomorphic.

Definition 4 Let $\mathcal{S}$ be a complex surface. We say that $\mathcal{S}$ is **real** if there exists a biholomorphism $\sigma : \mathcal{S} \to \overline{\mathcal{S}}$ such that σ^2 is the identity map.

As is often the case with Beauville surfaces, the above geometric condition can be translated into purely group theoretic terms.

Definition 5 Let G be a Beauville group. We say that G is **strongly real** if there exists a Beauville structure $X = \{\{x_1, y_1\}, \{x_2, y_2\}\}$ such that there exists an automorphism $\phi \in \mathrm{Aut}(G)$ and elements $g_i \in G$ for $i = 1, 2$ such that

$$g_i \phi(x_i) g_i^{-1} = x_i^{-1} \text{ and } g_i \phi(y_i) g_i^{-1} = y_i^{-1}$$

for $i = 1, 2$. In this case we also say that the Beauville structure X is a strongly real Beauville structure.

In practice we can always replace one generating pair by some conjugate of it and so we can take $g_1 = g_2 = e$ and often this is what is done in practice.

In [6] Bauer, Catanese and Grunewald show that a Beauville surface is real if, and only if, the corresponding Beauville group and structure are strongly real.

Example 6 In [15] Catanese classified the abelian Beauville groups by proving the following.

Theorem 7 *If G is an abelain group, then G is a Beauville group if, and only if, $G \cong \mathbb{Z}_n \times \mathbb{Z}_n$ where gcd(n,6)=1 and $\mathbb{Z}_n$ denotes the cyclic group of order $n > 1$.*

This theorem immediately gives us the following.

Corollary 8 *Every abelian Beauville group is a strongly real Beauville group making any Beauville structure for these groups strongly real.*

Proof If H is an abelian group, then the map $H \to H$, $x \mapsto -x$ is an automorphism. $\square$

More recent (and group theoretic) motivation comes from the following. The absolute Galois group $\mathrm{Gal}(\overline{\mathbb{Q}}/\mathbb{Q})$ is very poorly understood. Indeed, The Inverse Galois Problem—arguably the hardest open problem in algebra today—forms just one small part of efforts to understand $\mathrm{Gal}(\overline{\mathbb{Q}}/\mathbb{Q})$ (it amounts to showing that every finite group arises as the quotient of $\mathrm{Gal}(\overline{\mathbb{Q}}/\mathbb{Q})$ by a topologically closed normal subgroup). When confronted with the task of understanding a group it is natural to consider an action of the group on some set. The group $\mathrm{Gal}(\overline{\mathbb{Q}}/\mathbb{Q})$ acts on the set of Beauville surfaces thanks to Grothendieck's theory of Dessins d'enfants ("children's drawings"). See [27, Sect. 11] for a more detailed discussion of this and related matters.

Henceforth we shall use the standard ATLAS notation for group theoretic concepts (aside from occasional deviations to minimise confusion with geometric concepts) as described in some detail in the introductory sections of [16]. In particular, given two groups A and B we use the following notation.

- We write $A \times B$ for the direct product of A and B, that is, the group whose members are ordered pairs (a, b) with $a \in A$ and $b \in B$ such that for $(a, b), (a', b') \in A \times B$ we have the multiplication $(a, b)(a', b') = (aa', bb')$. Given a positive integer k we write A^k for the direct product of k copies of A.
- We write $A.B$ for the extension of A by B, that is, a group with a normal subgroup isomorphic to A whose quotient is B (such groups are not necessarily direct products—for instance $SL_2(5) = 2.PSL_2(5)$).
- We write $A : B$ for a semi-direct product of A and B, also known as a split extension A and B, that is, there is a homomorphism $\phi : B \to Aut(A)$ with elements of this group being ordered pairs (b, a) with $a \in A$ and $b \in B$ such that for $(b, a), (b', a') \in A : B$ we have the multiplication $(a, b)(a', b') = (bb', a^{\phi(b')}a')$.
- We write $A \wr B$ for the wreath product of A and B, that is, if B is a permutation group on n points then we have the split extension $A^n : B$ with B acting in a way that permutes the n copies of A.

In several places we shall refer to 'straightforward computations' or calculations that readers can easily reproduce for themselves. On these occasions either of Magma [10] or GAP [32] can easily be used to do this.

In Sect. 2 we will discuss the finite simple groups and in particular a conjecture of Bauer, Catanese and Grunewald concerning which of these groups are strongly real Beauville groups. In Sect. 3 our attention turns to the characteristically simple groups and in particular the recent work of Jones which we push further in the cases of the symmetric and alternating groups in Sect. 4. We go on in Sect. 5 to discuss which of the almost simple groups are strongly real Beauville groups. Finally, in Sect. 6 we briefly discuss nilpotent groups and p-groups.

2 The Finite Simple Groups

Naturally, a necessary condition for being a strongly real Beauville group is being a Beauville group. Furthermore, a necessary condition for being a Beauville group is being 2-generated: we say that a group G is 2-generated if there exist two elements $x, y \in G$ such that $\langle x, y \rangle = G$. It is an easy exercise for the reader to show that the alternating groups A_n for $n \geq 3$ are 2-generated. In [31] Steinberg proved that the simple groups of Lie type are 2-generated and in [1] Aschbacher and Guralnick showed that the sporadic simple groups are 2-generated. We thus have that all of the non-abelian finite simple groups are 2-generated making them natural candidates for Beauville groups. This lead Bauer, Catanese and Grunewald to conjecture that aside from A_5, which is easily seen to not be a Beauville group, every non-abelian finite simple group is a Beauville group—see [6, Conjecture 1] and [7, Conjecture 7.17]. This suspicion was later proved correct [19, 20, 24, 25], indeed the full theorem proved by the author, Magaard and Parker in [20] is actually a more general statement about quasisimple groups (recall that a group G is quasisimple if it is generated by its commutators and the quotient by its center $G/Z(G)$ is a simple group).

Having found that almost all of the non-abelian finite simple groups are Beauville groups, it is natural to ask which of the non-abelian finite simple groups are strongly real Beauville groups. In [6, Sect. 5.4] Bauer et al. wrote

> There are 18 finite simple nonabelian groups of order $\leq 15,000$. By computer calculations we have found strongly [real] Beauville structures on all of them with the exceptions of A_5, $PSL_2(7)$, A_6, A_7, $PSL_3(3)$, $U_3(3)$ and the Mathieu group M_{11}.

On the basis of these computations they conjectured that all but finitely many non-abelian finite simple groups are strongly real Beauville groups. Several authors have worked on this and many special cases are now known to be true.

- In [21] Fuertes and González-Diez showed that the alternating groups A_n ($n \geq 7$) and the symmetric groups S_n ($n \geq 5$) are strongly real Beauville groups by explicitly writing down permutations for their generators and the automorphisms and applying some of the classical theory of permutation groups to show that their elements had the properties they claimed. Subsequently the alternating group A_6 was also shown to be a strongly real Beauville group.
- In [22] Fuertes and Jones prove that the simple groups $PSL_2(q)$ for prime powers $q > 5$ and the quasisimple groups $SL_2(q)$ for prime powers $q > 5$ are strongly real Beauville groups. As with the alternating and symmetric groups, these results are proved by writing down explicit generators, this time combined with a celebrated theorem usually (but historically inaccurately) attributed to Dickson for the maximal subgroups of $PSL_2(q)$. General lemmas for lifting Beauville structures from a group to its covering groups are also used.
- Settling the case of the sporadic simple groups makes no impact on the above conjecture, there being only 26 of them. Nonetheless, for reasons we shall return to below, in [18] the author determined which of the sporadic simple groups are strongly real Beauville groups, including the '27th sporadic simple group', the Tits group $^2F_4(2)'$. Of all the sporadic simple groups only the Mathieu groups M_{11} and M_{23} are not strongly real. For all of the other sporadic groups smaller than the Baby Monster group $\mathbb{B}$ explicit words in the 'standard generators' [34] for a strongly real Beauville structure are given. (For those unfamiliar with standard generators, we will describe these in Sect. 5.) For the Baby Monster group $\mathbb{B}$ and Monster group $\mathbb{M}$ character theoretic methods are used.

As we can see from the above bullet points, several of the groups that Bauer, Catanese and Grunewald could not find strongly real Beauville structures for do indeed have strongly real Beauville structures. In particular, we note that the group $PSL_2(9) \cong A_6$ is in fact strongly real.

Using the results mentioned above, combined with unpublished calculations, the author has pushed Bauer, Catanese and Grunewald's original computations to every non-abelian finite simple group of order at most 100,000,000 and, as we noted above, several much larger ones in [18]. Many of the smaller groups seemed to require the use of outer automorphisms to make their Beauville structures strongly real, which explains much of the above difficulty in finding strongly real Beauville structures in certain groups. Slightly larger groups had enough conjugacy classes for inner automorphisms to be used instead. Consequently, it seems that 'small' non-abelian finite simple groups fail to be strongly real if they have too few conjugacy classes (as is the case with A_5 and as we would intuitively expect) or if they have no outer automorphisms—a phenomenon that is extremely rare. We are thus lead to the following somewhat stronger conjecture.

Conjecture 1 *All non-abelian finite simple groups apart from A_5, M_{11} and M_{23} are strongly real Beauville groups.*

To add further weight to this conjecture we verify this conjecture for the Suzuki groups $^2B_2(2^{2n+1})$. Let $q = 2^{2n+1}$.

Theorem 9 *Each of the groups $^2B_2(q)$ has a strongly real Beauville structure of type $((q - 1, q - 1, q - 1), (d_1, d_2, 2))$ where d_1 and d_2 are odd and coprime to $q - 1$.*

Throughout the following we shall be using the natural 4-dimensional representation of $^2B_2(q)$ over the field of order q as described in some detail in [35, Sect. 4.2]. To prove Thoerem 9 we will use knowledge of the maximal subgroups of the Suzuki groups. The following lemma was proved by Suzuki—see [35, Theorem 4.1]. Here we write E_q for the elementary abelian group of order q. Furthermore, by 'subfield subgroup' we mean either the subgroup $^2B_2(q_0)$ consisting of matrices whose entries come from a subfield of the field $\mathbb{F}_q$ of order $\mathbb{F}_{q_0}$ where $q_0 > 1$ divides q or one of its conjugates, those appearing in Lemma 10(v) being precisely the maximal subfield subgroups.

Lemma 10 *If $n > 1$, then the maximal subgroups of $^2B_2(q)$ are (up to conjugacy).*

  (i) $E_q.E_q : \mathbb{Z}_{q-1}$, *the subgroup of lower triangular matrices*
 (ii) $D_{2(q-1)}$
(iii) $\mathbb{Z}_{q+\sqrt{2q}+1} : 4$
(iv) $\mathbb{Z}_{q-\sqrt{2q}+1} : 4$
 (v) $^2B_2(q_0)$ *where $q = q_0^r$, r is prime and $q_0 > 2$.*

From the above the following can easily be deduced.

Lemma 11 *(a) If $x, y \in {}^2B_2(q)$ are two elements with the property that*

$$o(x) = o(y) = o(xy) = q - 1,$$

  then $\langle x, y \rangle = \mathbb{Z}_{q-1}$, $E_q.E_q : \mathbb{Z}_{q-1}$ or $^2B_2(q)$.
(b) If $x, y \in {}^2B_2(q)$ are two elements such that $o(x)$ and $o(y)$ are have orders dividing $q \pm \sqrt{2q} + 1$ and $o(xy) = 2$, then $\langle x, y \rangle = {}^2B_2(q)$ or a subfield subgroup.

Proof of Theorem 9 For our first generating pair we consider the following elements of $^2B_2(q)$ each of which are easily checked to have order 2 by direct calculation.

$$
t_1 := \begin{pmatrix} 0 & 0 & 0 & 1 \\ 0 & 0 & 1 & 0 \\ 0 & 1 & 0 & 0 \\ 1 & 0 & 0 & 0 \end{pmatrix}
\qquad
t_2 := \begin{pmatrix} 0 & 0 & 0 & \beta^{-1} \\ 0 & 0 & \beta^{-2^{n+1}+1} & 0 \\ 0 & \beta^{2^{n+1}-1} & 0 & 0 \\ \beta & 0 & 0 & 0 \end{pmatrix}
$$

$$
t_3 := \begin{pmatrix} 1 & 0 & 0 & 0 \\ 0 & 1 & 0 & 0 \\ \alpha^{2^{n+1}} & 0 & 1 & 0 \\ \alpha^2 & \alpha^{2^{n+1}} & 0 & 1 \end{pmatrix}
$$

where α and β are generators of the multiplicative group $\mathbb{F}_q^\times$. The element $x_1 = t_1 t_2$ has order $q - 1$. The characteristic polynomial of $y_1 = t_1 t_3$ is

$$
p_1(\lambda) = \lambda^4 + \alpha^2 \lambda^3 + \alpha^{2^{n+2}} \lambda^2 + \alpha^2 \lambda + 1
$$

and if we set $\gamma := \beta + \beta^{-1} + \beta^{2^{n+1}-1} + \beta^{1-2^{n+1}}$ then the characteristic polynomial of $t_1 t_2$ is

$$
p_2(\lambda) = \lambda^4 + \gamma \lambda^3 + (\beta^{2^{n+1}} + \beta^{-2^{n+1}+2} + \beta^{2^{n+1}-2} + \beta^{-2^{n+1}})\lambda^2 + \gamma \lambda + 1
$$

Comparing p_1 with p_2 we see that the two polynomials are equal if we have

$$
\gamma = \alpha^2 \text{ and }
$$

$$
\beta^{2^{n+1}} + \beta^{-2^{n+1}+2} + \beta^{2^{n+1}-2} + \beta^{-2^{n+1}} = \alpha^{2^{n+2}} = (\alpha^2)^{2^{n+1}} = \gamma^{2^{n+1}}.
$$

Since $a \mapsto a^2$ is an automorphism of our underlying field we see that the first of these equalities immediately implies the second if

$$
(\beta + \beta^{-1} + \beta^{2^{n+1}-1} + \beta^{1-2^{n+1}})^{2^{n+1}} = \beta^{2^{n+1}} + \beta^{-2^{n+1}+2} + \beta^{2^{n+1}-2} + \beta^{-2^{n+1}}.
$$

Since $\beta^{(2^{n+1}-1)2^{n+1}} = (\beta^{2^{2n+1}})^2 \beta^{-2^{n+1}} = \beta^{2-2^{n+1}}$ we can choose α and β to satisfy the above condition, so in particular we have that $t_1 t_2$ and $t_1 t_3$ have the same characteristic polynomial and thus both have order $q - 1$. Furthermore, these are both inverted by conjugation by t_1 since t_1, t_2 and t_3 all have order 2. Similarly we find that $t_1 t_2 t_1 t_3$ has characteristic polynomial of the correct form to have order $q - 1$. From Lemma 11(a) we see that these elements generate the group since x_1 and y_1 are not both contained in a cyclic subgroup (one of them is diagonal) and by direct calculation no one-dimensional subspace in the natural module is preserved by them so there is no proper subgroup containing each of these elements.

For the second triple we consider the matrices

$$x_2 := \begin{pmatrix} 0 & 0 & 0 & 1 \\ 0 & 0 & 1 & 0 \\ 0 & 1 & 0 & \delta^4 \\ 1 & 0 & \delta^4 & \delta^2 \end{pmatrix} \qquad y_2 := \begin{pmatrix} \epsilon^2 & \epsilon^4 & 0 & 1 \\ \epsilon^4 & 0 & 1 & 0 \\ 0 & 1 & 0 & 0 \\ 1 & 0 & 0 & 0 \end{pmatrix}$$

where $\delta, \epsilon \in \mathbb{F}_q$ are chosen so that $\delta \neq \epsilon$ and these do not have the correct form for these elements to have order $q - 1$. Direct calculation shows that these elements do not have orders 2 or 4 and that $o(x_2 y_2) = 2$. These elements must, therefore, have orders that divide $q \pm \sqrt{2q} + 1$. Furthermore their traces are ϵ^2 and δ^2 which can be chosen to be in no proper subfield since $x \mapsto x^2$ is an automorphism of the field $\mathbb{F}_q$. These elements must therefore generate the group by Lemma 11(b). Further direct calculation shows that $x_2^{t_1} = x_2^{-1}$ and $y_2^{t_1} = y_2^{-1}$. $\qquad\square$

3 Characteristically Simple Groups

Another class of finite groups that has recently been studied from the viewpoint of Beauville constructions, and seems like fertile ground for providing further examples of strongly real Beauville groups, are the characteristically simple groups that we define as follows (the definition commonly given is somewhat different from that below but in the case finite groups it is equivalent to this).

Definition 12 A finite group G is said to be **characteristically simple** if G is isomorphic to some direct product H^k where H is a finite simple groups.

For example, as we saw in Theorem 7, if $p > 3$ is prime then the abelian Beauville groups isomorphic to $\mathbb{Z}_p \times \mathbb{Z}_p$ are characteristically simple.

Characteristically simple Beauville groups have recently been investigated by Jones in [28, 29] where the following conjecture is discussed.

Conjecture 2 *Let G be a finite non-abelian characteristically simple group. Then G is a Beauville group if and only if it is a 2-generator group not isomorphic to A_5.*

In particular, the main results of [28, 29] verify this conjecture in the cases where H is any of the alternating groups; the linear groups $PSL_2(q)$ and $PSL_3(q)$; the unitary groups $PSU_3(q)$; the Suzuki groups $^2B_2(2^{2n+1})$; the small Ree groups $^2G_2(3^{2n+1})$ and the sporadic simple groups.

For large values of k, the group H^k will not be 2-generated despite the fact that H will be as discussed in Sect. 2. The values of k for which H^k is 2-generated can be surprisingly large. For example, a special case of the results alluded to in the previous paragraph is the somewhat amusing fact that

$$A_5 \times A_5 \times A_5 \times A_5 \times A_5 \times A_5 \times A_5 \times A_5 \times A_5 \times A_5 \times$$

$$A_5 \times A_5 \times A_5 \times A_5 \times A_5 \times A_5 \times A_5 \times A_5 \times A_5$$

is a Beauville group, despite the fact that A_5 itself is not a Beauville group.

In general, the full automorphism group of H^k will be the wreath product $\mathrm{Aut}(H) \wr S_k$ where S_k is the kth symmetric group acting on the product by permuting the groups H. This bounteous supply of automorphisms makes it likely that characteristically simple Beaville groups are in general strongly real.

Question 1 *Which characteristically simple Beauville groups are strongly real?*

As a more specific conjecture on these matters we assert the following.

Conjecture 3 *If H is a finite simple group of order greater than 3, then the group $H \times H$ is a strongly real Beauville group.*

Note that Corollary 8 tells us that this is true for all abelian characteristically simple Beauville groups. For the nonabelian characteristically simple Beauville groups this conjecture seems rather distant given that, at the time of writing, we have neither a solution to Conjecture 1 nor do we know if $H \times H$ for a simple group H is even a Beauville group, let alone a strongly real one. Anyone tempted to extend the above conjecture to the products of a larger number of copies of simple groups should see the remarks following Lemma 15, although some hope is provided by the results proven in Sect. 4.

Theorem 13 *Let G be a strongly real Beauville group with coprime strongly real Beauville structure $\{\{x_1, y_1\}, \{x_2, y_2\}\}$. Furthermore, suppose that there exists an automorphism $\phi \in \mathrm{Aut}(G)$ such that*

$$\phi(x_1) = x_1^{-1}, \ \phi(y_1) = y_1^{-1}, \ \phi(x_2) = x_2^{-1} \ and \ \phi(y_2) = y_2^{-1}.$$

Then the group $G \times G$ is a strongly real Beauville group.

Proof Consider the following elements of $G \times G$

$$g_1 = (x_1, x_2), h_1 = (y_1, y_2), g_2 = (x_2, x_1) \ and \ h_2 = (y_2, y_1).$$

The pair $\{g_1, h_1\}$ generate the whole of $G \times G$ since the elements $g_1^{o(x_2)}$ and $h_1^{o(y_2)}$ generate the first factor whilst the elements $g_1^{o(x_1)}$ and $h_1^{o(y_1)}$ generate the second factor thanks to our hypothesis that $o(x_1)o(x_2)o(x_1y_1)$ is coprime to $o(x_2)o(y_2)o(x_2y_2)$. Similarly $\langle g_2, h_2 \rangle = G \times G$.

We define an automorphism $\psi \in \mathrm{Aut}(G \times G)$ such that for every $(g, h) \in G \times G$

$$\psi(g, h) = (\phi(g), \phi(h)).$$

This automorphism clearly makes the above Beauville structure for $G \times G$ a strongly real Beauville structure. $\qquad\square$

Corollary 14 *Conjecture 3 is true for each of the following groups.*

(a) The alternating groups A_n for $n \geq 6$;
(b) The linear groups $PSL_2(q)$ for prime powers $q > 5$;
(c) The Suzuki groups ${}^2B_2(2^{2n+1})$;
(d) All simple groups of order at most $100,000,000$;
(e) The sporadic simple groups.

Proof For part (a) the results proved in Sect. 4 provide a strongly real Beauville structure for $A_n \times A_n$ for sufficiently large n, the smaller cases being straightforward calculations that are easily performed separately.

For part (b) we note that the strongly real Beauville structures constructed by Fuertes and Jones in [22] for the groups $PSL_2(q)$ satisfy the hypotheses of Theorem 13.

For part (c) we note that the strongly real Beauville structures for ${}^2B_2(2^{2n+1})$ we constructed in Theorem 9 are coprime and satisfy the hypotheses of Theorem 13.

For part (d) we note that the author's computations alluded to in Sect. 2 were performed in such a way that the hypotheses of Theorem 13 are satisfied.

Finally for part (e) we observe that for all the sporadic groups, apart from the Mathieu groups M_{11} and M_{23}, the structures given by the author in [18] satisfy the hypotheses of Theorem 13. The groups M_{11} and M_{23} are dealt with separately in Lemma 15. $\qquad\square$

We remark that the strongly real Beauville structures for the quasisimple groups $SL_2(q)$ where $q > 5$ constructed by Fuertes and Jones in [22] also satisfy the hypotheses of Theorem 13 and so the groups $SL_2(q) \times SL_2(q)$ are also strongly real.

Unfortunately, Theorem 13 cannot be applied to the strongly real Beauville structures constructed by Fuertes and González-Diez in [21] for the symmetric and alternating groups. This is because the types of the Beauville structures in [21] fail to satisfy the coprime hypothesis since their structures use several elements of order 2. We return to this point in Sect. 4.

Comparing the statement of Conjecture 1 with the statement of Conjecture 3 the reader should immediately be asking "what about the alternating group A_5 and the Matheiu groups M_{11} and M_{23}?" This concern is immediately addressed by the following further piece of evidence for Conjecture 3.

Lemma 15 *The groups $A_5 \times A_5$, $M_{11} \times M_{11}$ and $M_{23} \times M_{23}$ are strongly real Beauville groups.*

Proof This is a straightforward computational calculation. Consider the following permutations.

$$x_1 := (1, 2, 3, 4, 5)(6, 7, 8, 9, 10) \qquad y_1 := (2, 3, 4)(7, 10)(6, 9),$$
$$x_2 := (1, 4, 3, 2, 5)(7, 8, 9) \qquad\qquad y_2 := (1, 2)(4, 5)(6, 9, 8, 7, 10) \text{ and}$$
$$a := (1, 5)(2, 4)(6, 10)(7, 9)$$

The set $\{\{x_1, y_1\}, \{x_2, y_2\}\}$ gives a Beauville structure for the group $A_5 \times A_5$ of type $((5,6,5),(15,10,15))$ acting intransitively on $5 + 5$ points as a subgroup of the symmetric group S_{10}. The automorphism α defined by conjugation by a has the property that

$$\alpha(x_1) = x_1^{-1}, \alpha(y_1) = y_1^{-1}, \alpha(x_2) = x_2^{-1} \text{ and } \alpha(y_2) = y_2^{-1}$$

from which we have that this Beauville structure is strongly real.

Next, the group $M_{11} \times M_{11}$. Consider the following permutations.

$$x_1 := (1, 2, 3, 4, 5, 6, 7, 8, 9, 10, 11)(12, 13, 14, 15, 16, 17, 18, 19, 20, 21, 22)$$
$$y_1 := (1, 6, 10, 5, 2, 7, 4, 9, 11, 8, 3)(12, 14, 19, 16, 21, 18, 13, 17, 22, 20, 15)$$
$$x_2 := (1, 3, 9, 11, 10, 7, 2, 4)(5, 8)(12, 14, 20, 22, 19, 21, 16, 13)(15, 18)$$
$$y_2 := (2, 6, 9, 4, 8, 3, 7, 5)(10, 11)(12, 13)(14, 17, 21, 18, 16, 20, 15, 19)$$
$$a := (1, 22)(2, 21)(3, 20)(4, 19)(5, 18)(6, 17)(7, 16)(8, 15)(9, 14)(10, 13)(11, 12)$$

The set $\{\{x_1, y_1\}, \{x_2, y_2\}\}$ gives a Beauville structure for the group $M_{11} \times M_{11}$ of type $((11,11,11),(8,8,8))$ acting intransitively on $11+11$ points as a subgroup of the symmetric group S_{22}. The automorphism α defined by conjugation by a has the property that

$$\alpha(x_1) = x_1^{-1}, \alpha(y_1) = y_1^{-1}, \alpha(x_2) = x_2^{-1} \text{ and } \alpha(y_2) = y_2^{-1}$$

from which we have that this Beauville structure is strongly real.

Finally for $M_{23} \times M_{23}$ we similarly have that the permutations

$$x_1 := (1, 22, 5, 17, 6, 10, 18, 16, 19, 8, 9, 15, 13, 14, 21, 4, 3, 7, 23, 20, 2, 12, 11)$$
$$(24, 40, 44, 43, 26, 33, 34, 32, 38, 39, 28, 31, 29, 37, 41, 30, 42, 25, 46, 36,$$
$$35, 45, 27)$$
$$y_1 := (1, 16, 3, 14, 7, 15, 18, 22, 21, 8, 20, 10, 4, 17, 19, 13, 5, 6, 23, 9, 2, 12, 11)$$
$$(24, 41, 42, 34, 28, 30, 43, 37, 27, 39, 26, 25, 29, 32, 40, 33, 44, 31, 46, 36,$$
$$35, 45, 38)$$
$$x_2 := (1, 3, 19, 7, 18, 4, 11, 21, 16, 14, 6)(2, 23, 9, 17, 15, 20, 22, 10, 13, 12, 8)$$
$$(24, 45, 39, 35, 34, 37, 25, 27, 32, 30, 38)(26, 36, 43, 29, 40, 28, 44, 46, 41,$$
$$33, 31)$$

$$y_2 := (1, 6, 22, 9, 16, 17, 5, 19, 11, 18, 2)(3, 14, 13, 23, 4, 12, 15, 10, 7, 21, 8)$$
$$(24, 34, 33, 44, 39, 26, 40, 37, 32, 35, 43)(25, 41, 46, 45, 29, 36, 28, 42, 30,$$
$$31, 38)$$
$$a := (1, 46)(2, 45)(3, 44)(4, 43)(5, 42)(6, 41)(7, 40)(8, 39)(9, 38)(10, 37)$$
$$(11, 36)(12, 35)(13, 34)(14, 33)(15, 32)(16, 31)(17, 30)(18, 29)(19, 28)$$
$$(20, 27)(21, 26)(22, 25)(23, 24)$$

define a strongly real Beauville structure of type $((23,23,23),(11,11,11))$ for the group $M_{23} \times M_{23}$ acting intransitively on $23 + 23$ points as a subgroup of the symmetric group S_{46}. $\square$

We remark that in the examples of the above lemma, the automorphisms used are outer automorphisms that interchange the two factors. The lack of automorphisms that stop both M_{11} and M_{23} being strongly real will therefore also stop the groups $M_{11} \times M_{11} \times M_{11}$ and $M_{23} \times M_{23} \times M_{23}$ being strongly real Beauville groups. Furthermore it is easy to see that the permutations given in the proof of Lemma 15 can be adapted to construct a strongly real Beauville structure of type $((88,88,88),(88,88,88))$ for the group $M_{11} \times M_{11} \times M_{11} \times M_{11}$ and to construct a strongly real Beauville structure of type $((253,253,253),(253,253,253))$ for the group $M_{23} \times M_{23} \times M_{23} \times M_{23}$. Similarly $A_5 \times A_5 \times A_5$ is not a strongly real Beauville group. It follows that any extension of Conjecture 3 to products of a larger number of copies of simple groups will necessarily have a much more complicated statement. It is likely that similar remarks apply to M_{11}^{2k+1}, M_{11}^{2k}, M_{23}^{2k+1}, M_{23}^{2k}, A_5^{2k+1} and A_5^{2k} for small values of k.

In light of the above it is natural to ask the following.

Question 2 *Let H be a finite simple group, $n \in \mathbb{Z}^+$ and $G = H^n$. When are inner automorphism sufficient to make G strongly real and when are outer automorphisms interchanging the factors required? Moreover does this have any geometric significance for the corresponding surfaces?*

4 The Symmetric and Alternating Groups

In the last section we discussed characteristically simple groups of the form $H \times H$ for some simple group H. In this section we prove slightly stronger results in the case of the alternating groups and a related result for the symmetric groups. In each of the below results conjugacy of elements is taken care of by the well known fact that two elements of the symmetric group are conjugate if, and only if, they have the same cycle type. We will use the following recent results of Jones.

Lemma 16 *Let $H \leq S_n$.*

(a) If H is primitive and contains a cycle that fixes at least three points then $H \geq A_n$.
(b) If H is transitive and contains an m-cycle where $m > n/2$ and m is coprime to n then H is primitive.

(c) If H is primitive and contains a cycle fixing two points then either $H \geq A_n$ or $PGL_2(q) \leq H \leq P\Gamma L_2(q)$ with $n = q + 1$ for some prime power q.

Proof See [30] and [28, Sect. 6]. $\qquad\square$

Before proving our main results we recall some facts about generating pairs in simple and characteristically simple groups. Let H be a finite simple group. In [26] Philip Hall showed that the largest k such that the characteristically simple group H^k is 2-generated is equal to the number of orbits of $Aut(H)$ on generating pairs of H. (He proved similar results for more general n-tuples but we will not be needing these results here.) To show that H^k for some k is generated by a pair of elements it is sufficient to show that each of the 'coordinates' of these elements (that is, the parts of each permutation that correspond to each of the factors) are inequivalent under the action of $Aut(H)$. For $n \neq 1, 2, 6$ we have that $Aut(A_n) = S_n$ and in the case $n = 6$ the symmetric group S_6 is an index 2 subgroup of $Aut(A_6) \cong P\Gamma L_2(9)$.

Lemma 17 *Let $n \geq 11$ be odd and let $k \leq (n - 6)/2$ be positive integers. Then A_n^k is a strongly real Beauville group.*

Proof Since A_n is simple for every $n > 5$ we have that, by the remarks of the previous paragraph, it is sufficient to find k pairs of generating pairs $T_{ij} = \{x_{ij}, y_{ij}\}$ $i = 1, 2$, $j = 1, \ldots, k$ such that for a fixed i no T_{ij} is an image of $T_{ij'}$ under the actions of automorphisms of A_n for distinct j and j'.

For our first pairs we set

$$x_{1j} = (1, \ldots, 2j + 3) \text{ and } y_{1j} = (2j + 3, \ldots, n)$$

for $1 \leq j < (n-6)/4$. These are cycles of odd length and are thus even permutations. Their product is an n-cycle. It is easy to check that $\langle x_{1j}, y_{1j} \rangle$ is primitive and thus equal to A_n since the group contains a cycle with at least three fixed points by Lemma 16(a). These elements are both inverted by the automorphism defined by conjugation by

$$t = (1, 2j)(2, 2j - 1) \cdots (j, j + 1)(2j + 2, n)(2j + 3, n - 1) \cdots$$
$$((2j + n + 1)/2, (2j + n + 3)/2)$$

which has only one fixed point, namely $2j + 1$.

For our second pairs we consider the permutations

$$x_{2j} = (1, \ldots, n - 2) \text{ and } y_{2j} = (j + 1, j + 2)(n - j - 1, n - j - 2)$$
$$\times ((n - 1)/2, n - 1)((n + 1)/2, n)$$

for $1 \leq j < (n - 5)/2$. These are again both even permutations, their product this time being an $(n - 2)$-cycle. To confirm that these elements generate the group we note that $\langle x_{2j}, y_{2j} \rangle$ is clearly transitive and so by Lemma 16(b) must be primitive. It now follows from Lemma 16(c) that $\langle x_{2j}, y_{2j} \rangle = A_n$ since $n > 9$ and the only elements of order 2 in $P\Gamma L_2(q)$ do not have the same cycle type as y_{2j}. (In the case

$n = 9$ these permutations generate $PSL_2(8)$.) These elements are both inverted by the automorphism defined by conjugation by

$$(2, n-2)(3, n-3)\cdots((n-1)/2, (n+1)/2)(n-1, n)$$

which has only one fixed point, namely 1, and thus differs from t solely by an inner automorphism. $\qquad\square$

Lemma 18 *Let $n \geq 12$ be an even integer and let $k \leq (n-8)/4$. Then A_n^k is a strongly real Beauville group.*

Proof Again, we seek a collection of generating pairs that are not mapped to one another by automorphisms of A_n.

For our first pairs we consider the following elements.

$$x_{1j} = (1, \ldots, 2j+5) \text{ and } y_{1j} = (n, \ldots, 2j+5, 2j+4)$$

for $1 \leq j \leq (n-8)/4$. These are cycles of odd length and are thus even permutations. Their product is an $(n-1)$-cycle. It is easy to check that $\langle x_{1j}, y_{1j}\rangle$ is primitive and thus equal to A_n since the group contains a cycle with at least three fixed points by Lemma 16(a). These elements are both inverted by the automorphism defined by conjugation by

$$t = (2j+4, 2j+5)(1, 2j+3)\cdots(j+1, j+3)(2j+6, n)\cdots$$
$$((n+2j+4)/2 - 1, (n+2j+4)/2 + 1)$$

which has precisely two fixed points namely $j+2$ and $(n+2j+4)/2$.

For our second pair of generators we consider the permutations

$$x_{2j} = (1, \ldots, n-2)(n-1, n) \text{ and}$$

$$y_{2j} = (n/2, (n-2)/2, n-1)(j+1, j, n, n-j-1, n-j-2)$$

for $1 \leq j < (n-2)/2$. These are both even permutations and their product is an $(n-3)$-cycle that fixes the points $j, n-j-2$ and $(n-2)/2$. The subgroup $\langle x_{2j}y_{2j}, y_{2j}^3\rangle$ fixes the point $(n-2)/2$ and is transitive on the remaining $n-1$ points. It follows that the group $\langle x_{2j}, y_{2j}\rangle$ is 2-transitive and is therefore primitive. Since the group $\langle x_{2j}, y_{2j}\rangle$ also contains the 3-cycle y_{2j}^5 (and the 5-cycle y_{2j}^3) it is equal to A_n by Lemma 16(a). These elements are both inverted by the automorphism defined by conjugation by

$$(1, n-2)(2, n-3)\cdots((n-2)/2, n/2)$$

which has precisely two fixed points, namely $n-1$ and n and thus differs from t solely in an inner automorphism. $\qquad\square$

When considering the alternating groups it is natural to seek similar results for the symmetric groups. Unfortunately, here we are somewhat limited: if $k > 2$ then S_n^k is not 2-generated since its abelianisation is $\mathbb{Z}_2^k$ and this is not 2-generated. It follows that we can only find analogous results for $k \leq 2$ and since $k = 1$ comes straight from the work of Fuertes and González-Diez we are left only to consider the case $k = 2$. Note that in this case the outer automorphism of $S_n \times S_n$ that interchanges the two factors is useless since the only permutations that are inverted by this automorphism are even.

Lemma 19 *For $n \geq 5$ the group $S_n \times S_n$ is strongly real.*

Proof We will explicitly construct our Beauville structure in the case of n even and then describe the differences in the case of n odd.

For our first pair we consider the following elements.

$$x_1 = (1, \ldots, n-1)(2n-1, 2n) \text{ and } y_1 = (n, n-1)(n+1, \ldots, 2n-1)$$

The product of these permutations is a pair of n-cycles. It is easy to check that $\{x_1^2, y_1^{n-1}\}$ generates the first of the two factors whilst $\{x_1^{n-1}, y_1^2\}$ generates the second and so $\langle x_1, y_1 \rangle$ is the whole group. These elements are both inverted by the automorphism defined by conjugation by

$$t = (1, n-2)(2, n-3) \cdots (n/2, n/2 - 1)(n+1, 2n-2) \cdots (3n/2, 3n/2 - 1)$$

which has precisely four fixed points, namely $n-1$, n, $2n-1$ and $2n$.

For our second pair of generators we consider the permutations

$$x_2 = (1, 2, 3)(4, \ldots, n)(n+4, n+3, n+2, n+1)$$

and

$$y_2 = (4, 3, 2, 1)(n+1, n+2, n+3)(n+4, \ldots, 2n)$$

The product of these permutations is a pair of $n-2$ cycles. It is easy to check that $\{x_1^4, y_1^{3(n-3)}\}$ generates the first factor (this group is easily seen to be 2-transitive, and thus primitive, by considering conjugates of $y_1^{3(n-1)}$ under powers of x_1^4 and since this group contains a 4-cycle it contains the whole of S_n by [30, Corollary 1.3]). Similarly $\{x_1^{3(n-1)}, y_1^4\}$ generates the second factor. These elements are both inverted by the automorphism defined by conjugation by

$$(1, 3)(5, n) \cdots (n/2+2, n/2+3)(n+1, n+3)(n+5, 2n) \cdots (3n/2+2, n/2+3)$$

which has precisely four fixed points, namely 2, 4, $n+2$ and $n+4$ and thus differs from t solely in an inner automorphism. (The case $n = 6$ requires a little care—using

$$x_2 = (1, 2, 3, 4)(10, 11, 12), y_2 = (4, 5, 6)(7, 8, 9, 10),$$

and the automorphism defined by $(1, 3)(5, 6)(7, 9)(11, 12)$ avoids being trapped inside copies of S_5 acting transitively on six points.)

If n is odd then for the first pair we need only replace the $(n-1)$-cycles with n-cycles to ensure that the elements have odd parity and for the automorphism instead use

$$t = (n, n-1)(n-1, 1) \cdots ((n-1)/2 - 1, (n-1)/2 + 1) \cdots$$

$$(2n, 2n-1)(2n-2, n+1) \cdots ((3n-1)/2 - 1, (3n-1)/2 + 1).$$

For the second generating pair we must now replace the 4-cycles with some longer p-cycle whose length is coprime to $3(n-3)$ (if 5 fails then $n \geq 2 \times 5 + 3$ and we can try 7; if both 5 and 7 fail then $n \geq 2 \times 5 \times 7 + 3$ and we can try 11 etc.). The permutation of order 2 for the automorphism needs to be adjusted in the obvious manner (i.e. $(1, 3)(4, p)(5, p-1) \cdots (p+1, n)(p+2, n-1) \cdots$). Again, the smallest case needs separate attention but it is easily checked that if

$$x_1 = (1, 4)(2, 5)(6, 10)(7, 8, 9) \text{ and } y_1 = (1, 5)(2, 3, 4)(6, 9)(7, 10)$$

and

$$x_2 = (1, 2, 3, 4, 5)(6, 7, 9, 10) \text{ and } y_2 = (5, 4, 2, 1)(10, 9, 8, 7, 6)$$

then $\{\{x_1, y_1\}, \{x_2, y_2\}\}$ is a strongly real Beauville structure whose elements are inverted by conjugation by the element $(1, 5)(2, 4)(6, 10)(7, 9)$. $\qquad\qquad\square$

5 Almost Simple Groups

Let G be a group. Recall that we say G is almost simple if there exists a simple group S such that $S \leq G \leq \mathrm{Aut}(S)$. For example, any simple group is almost simple, as are the symmetric groups. Given our earlier remarks on the finite simple groups it is natural to ask the following.

Question 3 *Which of the almost simple groups are strongly real?*

This is particularly pertinent in light of Fuertes and González-Diez proof that the symmetric groups S_n for $n > 5$ are strongly real. Unfortunately, the general picture here is much more complicated with many almost simple groups not even being Beauville groups, let alone strongly real Beauville groups. Worse, infinitely many of the almost simple groups are not even 2-generated: the smallest example is $PSL_4(9)$ whose outer automorphism group is $\mathbb{Z}_2 \times D_8$ (and more generally, if p is an odd prime and r is an even positive integer then $Aut(PSL_4(p^r))$ is not 2-generated). We can at least add the following to the list.

Theorem 20 *The non-simple almost simple sporadic groups are strongly real Beauville groups.*

Before proceeding to the proof of Theorem 20 we make the following remarks for those who are unfamiliar with standard generators of finite groups (those who are familiar with them may skip to the proof in the next paragraph). Any given group will have many generating sets and in particular if $x, y \in G$ are such that $\langle x, y \rangle = G$ then $\langle x^g, y^g \rangle = G$ for any $g \in G$. To provide some standardisation to computational group theory, Wilson [34] introduced the notion of 'standard generators'. These are generators for a group that are unusually easy to find and are specified in terms of which conjugacy classes that they belong to (which can often be determined solely from their orders) and which classes some word(s) in these elements belong to. Representatives for many of the finite simple groups and various other groups closely related to them may be found in explicit permutations and/or matrices for many of their most useful representations on the web-based Atlas of Group Representations [33].

To construct our Beauville structures that prove Theorem 20 we proceed as follows. We first recall some well-known facts about the sporadic simple groups. If G is one of the 27 sporadic simple groups (including the Tits group ${}^2F_4(2)'$) then the outer automorphism group of G has order at most 2 and that in all cases in which there exists a non-trivial outer automorphism $Aut(G)$ is a non-split extension, apart from the Tits group ${}^2F_4(2)'$ and may thus be written $G : 2$ in ATLAS notation (see Sect. 1). Let G be a simple group such that $Aut(G) = G : 2$. Let $t, t' \in G : 2$ have order 2 such that one of the elements lies in G and the other lies in $G : 2 \setminus G$. For $i = 1, 2$ we define the elements $x_i = tt'^{g_i}$ for some $g_i \in G : 2$. If for $i = 1, 2$ $u(i) \in C_G(t)$, then we can further define the elements $y_i = (x_i^{j(i)})^{u(i)}$ for some positive integers $j(i)$. Note that since $u(i)$ commutes with t the automorphism defined by conjugation by t inverts both x_i and y_i. Using knowledge of the subgroup structure of $G : 2$ it is often possible to choose the elements g_1, g_2, $u(1)$ and $u(2)$ in such a way that $\langle x_1, y_1 \rangle = \langle x_2, y_2 \rangle = G : 2$. Unfortunately, in the case of almost simple groups we must have that the orders of x_i and y_i all have even order and so verifying the conjugacy condition (†) of Definition 2 is more difficult than simply showing that $o(x_1)o(y_1)o(x_1y_1)$ is coprime to $o(x_2)o(y_2)o(x_2y_2)$. For some of the larger groups verifying that they generate the whole group can also be difficult. In these cases, generation is verified by finding words in our elements with the property that no proper subgroup can contain them (in many cases the maximal subgroups for these groups may be found in [16]). The words defining our Beauville structures are given in Table 1 and their types are given in Table 2.

We remark that this construction will not work in cases that are non-split extensions. This includes the almost simple 'sporadic' Tits group ${}^2F_4(2)$. Straightforward computations verify that this group is not a strongly real Beauville group.

In light of the above we make the following tentative conjecture.

Conjecture 4 *A split extension of a simple group is a Beauville group if, and only if, it is a strongly real Beauville group.*

Table 1 Words in the standard generators providing strongly real Beauville structures for each of the non-simple almost simple sporadic groups

G	t_1	t_2	x_1	x_2
$M_{12}:2$	c	$(cd)^6$	$t_1 t_2$	$t_1 t_2^d$
$M_{22}:2$	$((cd)^2 d)^5$	d^2	$t_1 t_2$	$t_1 t_2^c$
$J_2:2$	c	$(cd^2(cd)^2)^6$	$t_1 t_2$	$t_1 t_2^{d^4}$
HS: 2	c	$((cd)^3 cd^2)^5$	$t_1 t_2$	$t_1 t_2^d$
$J_3:2$	c	$(cd)^{12}$	$t_1 t_2$	$t_1 t_2^d$
McL: 2	c	$((cd)^2(cd^2)^2(cd)^2 d)^2$	$t_1 t_2$	$t_1 t_2^{(dcd)^2}$
He: 2	c	d^3	$t_1 t_2$	$t_1 t_2^{cd(cd^2)^2 c}$
Suz: 2	c	$(cd)^{14}$	$t_1 t_2$	$t_1 t_2^{(dc)^2(d^2 c)^2 d^2}$
O'N: 2	c	d^2	$t_1 t_2$	$t_1 t_2^{cd}$
$Fi_{22}:2$	$(cd^4)^{10}$	$(cd^3)^{15}$	$t_1 t_2^{dcd^6}$	$t_1 t_2^{dcd}$
HN: 2	c	$(cd^3(cd)^2)^{12}$	$t_1 t_2^{(dcd)^2 d^2}$	$t_1 t_2^{dcd^4 cd^2}$
Fi_{24}	d^4	$((cd)^2 d^3)^3 3$	$t_1 t_2^{d^4 c}$	$t_1 t_2^{dcd^2 c}$

G	u_1	u_2	$j(1)$	$j(2)$
$M_{12}:2$	$[c,(dc)^2 d^2]^3$	$(dc)^2 d[c,(dc)^2 d]^2$	1	1
$M_{22}:2$	$cd^2 cd[t_1, cd^2 cd]^5$	$[t_1, c]^2$	5	9
$J_2:2$	$d[c,d]^3$	$d[c,d]^3$	1	9
HS: 2	$d[c,d]$	$d[c,d]$	1	1
$J_3:2$	$d[c,d]^4$	$d[c,d]^4$	21	1
McL: 2	$d^2[c,d^2]^7$	$dcd[c,dcd]^7$	1	7
He: 2	$d[c,d]^7$	$d[c,d]^7$	15	19
Suz: 2	$d[c,d]^3$	$d[c,d]^3$	9	3
O'N: 2	$[c,d]^5$	$[c,d]^5$	7	1
$Fi_{22}:2$	$[t_1, d^3]^3$	$[t_1, d]^3$	3	1
HN: 2	$d^2[cd^2]^1 0$	$d[c,d]^4$	1	39
Fi_{24}	$c[t_1, c]$	$c[t_1, c]$	7	25

We remark that some (unpublished) progress on this conjecture has been made by the author's PhD student, Emilio Pierro, whilst the question of which of the groups $PGL_2(q)$ are Beauville is discussed by Garion in [23].

Table 2 The types of the Beauville structures specified by the words in Table 1

G	Type	G	Type
$M_{12}:2$	$((4,4,5),(6,6,3))$	He : 2	$((16,16,7),(30,30,5))$
$M_{22}:2$	$((12,12,4),(10,10,5))$	Suz : 2	$((10,10,3),(8,8,13))$
$J_2:2$	$((24,24,15),(14,14,7))$	O'N : 2	$((38,38,19),(56,56,28))$
HS : 2	$((8,8,8),(6,6,15))$	$Fi_{22}:2$	$((10,10,11),(12,12,4))$
$J_3:2$	$((34,34,17),(24,24,4))$	HN : 2	$((18,18,25),(44,44,22))$
McL : 2	$((8,8,3),(10,10,5))$	Fi_{24}	$((66,66,33),(84,84,26))$

6 Nilpotent Groups

It is immediate that the direct product of two Beauville groups of coprime order is again a Beauville group (though slightly more is true—see [4, Lemma 1.3]). Recall that a finite group is nilpotent if and only if it is the direct product of its Sylow subgroups. Since Sylow subgroups for different primes will have coprime orders this observation reduces the study of nilpotent Beauville groups to that of Beauville p-groups.

There is another motivation for wanting to study Beauville p-groups and that is to study how finite groups in general behave from the point of view of Beauville constructions. We saw in Sect. 2 that among the non-abelian finite simple groups only one fails to be a Beauville group and of the rest only two fail to be strongly real. This immediately raises the following question.

Question 4 *Are most Beauville groups strongly real Beauville groups?*

Finite simple groups are rare gems in the rough—for every positive integer n there are at most two finite simple groups of order n and for most values of n there are none at all. Taking our lead from their behaviour is therefore somewhat dangerous.

Few mathematicians outside finite group theory seem to realise that in some sense most finite groups are p-groups, indeed most finite groups are 2-groups. There are $49,910,529,484$ groups of order at most $2,000$. Of these $49,487,365,422$ have order precisely 1024—that's more than $99 \cdot 1\%$ of the total! When we throw in the other 2-groups of order at most 1 024 and the other p-groups of order at most $2,000$ we have almost all of them. Determining which of the Beauville p-groups are strongly real Beauville p-groups thus goes a long way to answering the above question for groups in general. (For details of these extraordinary computational feats and a historical discussion of the problem of enumerating groups of small order, which has been worked on for almost a century and a half, see the work of Besche et al. in [8, 9].)

Theorem 7 and Corollary 8 tell us that if $p \geq 5$ is prime then there are infinitely many strongly real Beauville p-groups—just let n be any power of p. These results are, however, useless for the primes 2 and 3. As far as the author is aware there are no known examples.

Problem 1 *Find strongly real Beauville 2-groups and 3-groups.*

The only known infinite family of Beauville 2-groups are those recently constructed by Barker, Boston, Peyerimhoff and Vdovina in [2]. One of the main results of [2] is that the groups constructed there are not strongly real. Furthermore there remain only finitely many known examples of Beauville 3-groups.

In general, p-groups have large outer automorphism groups [12, 13], so it seems likely that most Beauville p-groups are in fact strongly real. Again, as far as the author is aware, this matter remains largely uninvestigated.

Problem 2 *Find non-abelian strongly real Beauville p-groups.*

The best general discussion of work on Beauville p-groups is Boston's contribution to these proceedings [11]. The work of the Barker, Boston and the author in [4] and the work of Barker et al. in [2, 3, 5] are also worth consulting.

Acknowledgments The author wishes to express his deepest gratitude to the organisers of the Beauville Surfaces and Groups 2012 conference held in the University of Newcastle without which this volume, and thus the opportunity to present these results here, would not have been possible. The author also wishes to thank Professor Gareth Jones for many invaluable comments on earlier drafts of this article, particularly regarding the results concerned with products of symmetric and alternating groups. Finally, the author wishes to thank the anonymous referee whose comments and suggestions have substantially improved the readability of this paper, particularly bearing in mind the wide breadth of the audience for this work.

References

1. M. Aschbacher, R. Guralnick, Some applications of the first cohomology group. J. Algebra **90**(2), 446–460 (1984)
2. N. Barker, N. Boston, N. Peyerimhoff, A. Vdovina, An infinite family of 2-groups with mixed Beauville structures, to appear in Int. Math. Res. Not. arXiv:1304.4480
3. N. Barker, N. Boston, N. Peyerimhoff, A. Vdovina, Regular algebraic surfaces isogenous to a higher product constructed from group representations using projective planes, preprint (2011). arXiv:1109.6053
4. N.W. Barker, N. Boston, B.T. Fairbairn, A note on Beauville p-groups. Exp. Math. **21**(3), 298–306 (2012)
5. N. Barker, N. Boston, N. Peyerimhoff, A. Vdovina, New examples of Beauville surfaces. Monatsh. Math. **166**(3–4), 319–327 (2012). doi:10.1007/s00605-011-0284-6
6. I. Bauer, F. Catanese, F. Grunewald, Beauville surfaces without real structures, *Geometric Methods in Algebra and Number Theory*, vol. 235, Progress in Mathematics (Birkhuser, Boston, 2005)
7. I. Bauer, F. Catanese, F. Grunewald, Chebycheff and Belyi polynomials, Dessins d'Enfants, Beauville surfaces and group theory. Mediterr. J. Math. **3**, 121–146 (2006)
8. H.U. Besche, B. Eick, E.A. O'Brien, The groups of order at most 2 000. Electron. Res. Announc. Am. Math. Soc. **7**, 1–4 (2001)
9. H.U. Besche, B. Eick, E.A. O'Brien, A millennium project: constructing small groups. Int. J. Algebra Comput. **12**(5), 623–644 (2002)

10. W. Bosma, J. Cannon, C. Playoust, The Magma algebra system. I. The user language. J. Symb. Comput. **24**, 235–265 (1997)
11. N. Boston, A survey of Beauville p-groups, in *Proceedings of Conference on Beauville Surfaces and Group*, ed. by I. Bauer, S. Garion, A. Vdovina (Newcastle, 2012)
12. N. Boston, M.R. Bush, F. Hajir, Heuristics for p-class towers of imaginary quadratic fields, preprint (2011). arXiv:1111.4679v1
13. N. Boston, Embedding 2-groups in groups generated by involutions. J. Algebra **300**(1), 73–76 (2006)
14. F. Catanese, Fibered surfaces, varieties isogenous to a product and related moduli spaces. Am. J. Math. **122**(1), 1–44 (2000)
15. F. Catanese, Moduli spaces of surfaces and real structures. Ann. Math. **158**(2), 577–592 (2003)
16. J.H. Conway, R.T. Curtis, S.P. Norton, R.A. Parker, R.A. Wilson, *An ATLAS of Finite Groups* (Clarendon Press, Oxford, 1985)
17. J.D. Dixon, B. Mortimer, *Permutation Groups* (Springer, New York, 1996)
18. B.T. Fairbairn, Some exceptional Beauville structures. J. Group Theory **15**(5), 631–639 (2012). arXiv:1007.5050
19. B.T. Fairbairn, K. Magaard, C.W. Parker, Corrigendum to Generation of finite simple groups with an application to groups acting on Beauville surfaces, to appear in the Proc. Lond. Math. Soc
20. B.T. Fairbairn, K. Magaard, C.W. Parker, Generation of finite simple groups with an application to groups acting on Beauville surfaces. Proc. Lond. Math. Soc. **107**(5), 1220 (2013). doi:10.1112/plms/pdt037
21. Y. Fuertes, G. González-Diez, On Beauville structures on the groups S_n and A_n. Math. Z. **264**(4), 959–968 (2010)
22. Y. Fuertes, G.A. Jones, Beauville surfaces and finite groups. J. Algebra **340**, 13–27 (2011)
23. S. Garion, On Beauville structures for $PSL(2, q)$, preprint (2010). arXiv:1003.2792
24. S. Garion, M. Larsen, A. Lubotzky, Beauville surfaces and finite simple groups. J. Reine Angew. Math. **666**, 225–243 (2012)
25. R. Guralnick, G. Malle, Simple groups admit Beauville structures. J. Lond. Math. Soc. (2) **85**(3), 694–721 (2012)
26. P. Hall, The Eulerian functions of a group. Q. J. Math. **7**, 134–151 (1936)
27. G.A. Jones, Beauville surfaces and groups: a survey, to appear Fields Inst. Commun
28. G.A. Jones, Characteristically simple Beauville groups I: Cartesian powers of alternating groups. preprint (2013). arXiv:1304.5444
29. G.A. Jones, Characteristically simple Beauville groups II: low rank and sporadic groups. preprint (2013). arXiv:1304.5450
30. G.A. Jones, Primitive permutation groups containing a cycle, preprint (2012). arXiv:1209.5169v1
31. R. Steinberg, Generators for simple groups. Can. J. Math. **14**, 277–283 (1962)
32. The GAP Group, GAP—Groups, Algorithms, and Programming, Version 4.5.7; 2012. http://www.gap-system.org
33. R.A. Wilson et al., Atlas of finite group representations, version 3. http://brauer.maths.qmul.ac.uk/Atlas/ (2005) onwards
34. R.A. Wilson, Standard generators for sporadic simple groups. J. Algebra **184**(2), 505–515 (1996)
35. R.A. Wilson, *The Finite Simple Groups*, vol. 251, Graduate Texts in Mathematics (Springer, London, 2009)

Beauville Surfaces and Probabilistic Group Theory

Shelly Garion

Abstract A Beauville surface is a complex algebraic surface that can be presented as a quotient of a product of two curves by a suitable action of a finite group. Bauer, Catanese and Grunewald have been able to intrinsically characterize the groups appearing in minimal presentations of Beauville surfaces in terms of the existence of a so-called "Beauville structure". They conjectured that all finite simple groups, except A_5, admit such a structure. This conjecture has recently been proved by Guralnick-Malle and Fairbairn-Magaard-Parker. In this survey we demonstrate another approach towards the proof of this conjecture, based on probabilistic group-theoretical methods, by describing the following three works. The first is the work of Garion, Larsen and Lubotzky, showing that the above conjecture holds for almost all finite simple groups of Lie type. The second is the work of Garion and Penegini on Beauville structures of alternating groups, based on results of Liebeck and Shalev, and the third is the case of the group $\mathrm{PSL}_2(p^e)$, in which we give bounds on the probability of generating a Beauville structure. We also discuss other related problems regarding finite simple quotients of hyperbolic triangle groups and present some open questions and conjectures.

2000 Mathematics Subject Classification: 20D06 · 20H10 · 14J10 · 14J29 · 30F99

1 Beauville Surfaces and Beauville Structures

A *Beauville surface* S (over $\mathbb{C}$) is a particular kind of surface isogenous to a *higher product of curves*, i.e., $S = (C_1 \times C_2)/G$ is a quotient of a product of two smooth curves C_1 and C_2 of genus at least two, modulo a free action of a finite group G

The author was supported by the SFB 878 "Groups, Geometry and Actions".

S. Garion (✉)
Fachbereich Mathematik und Informatik, Universität Münster, Einsteinstrasse 62,
48149 Münster, Germany
e-mail: shelly.garion@uni-muenster.de

© Springer International Publishing Switzerland 2015
I. Bauer et al. (eds.), *Beauville Surfaces and Groups*, Springer Proceedings
in Mathematics & Statistics 123, DOI 10.1007/978-3-319-13862-6_5

which acts faithfully on each curve. For Beauville surfaces the quotients C_i/G are isomorphic to $\mathbb{P}^1$ and both projections $C_i \to C_i/G \cong \mathbb{P}^1$ are coverings branched over three points. A Beauville surface is in particular a minimal surface of general type.

Beauville [4] constructed a minimal surface of general type S with $K_S^2 = 8$ and $p_g = q = 0$ in the following way: take two curves $C_1 = C_2$ given by the Fermat equation $x^5 + y^5 + z^5 = 0$ and G the group $(\mathbb{Z}/5\mathbb{Z})^2$ acting on $C_1 \times C_2$ by

$$(a, b) \cdot ([x : y : z], [u : v : w]) = ([\xi^a x : \xi^b y : z], [\xi^{a+3b} u : \xi^{2a+4b} v : w]),$$

where $\xi = e^{\frac{2\pi i}{5}}$ and $a, b \in \mathbb{Z}/5\mathbb{Z}$. Then define S by the quotient $(C_1 \times C_2)/G$. Moreover $C_i \to C_i/G \cong \mathbb{P}^1_{\mathbb{C}}$ and both covers are branched in exactly three points. Curves with such properties are said to be *triangle curves*.

Inspired by this construction Catanese [5] observed that in general if C_1 and C_2 are two triangle curves with group G, if the action of G on the product $C_1 \times C_2$ is free, then $S = (C_1 \times C_2)/G$ is a *strongly rigid surface*, i.e., if S' is another surface homotopically equivalent to S then S' is either biholomorphic or antibiholomorphic to S. He proposed to name these surfaces *Beauville surfaces*.

A Beauville surface S is either of *mixed* or *unmixed* type according respectively as the action of G exchanges the two factors (and then C_1 and C_2 are isomorphic) or G acts diagonally on the product $C_1 \times C_2$. The subgroup G_0 (of index ≤ 2) of G which preserves the ordered pair (C_1, C_2) is then respectively of index 2 or 1 in G. Any Beauville surface S can be presented in such a way that the subgroup G_0 of G acts effectively on each of the factors C_1 and C_2. Catanese called such a presentation *minimal* and proved its uniqueness in [5]. In this survey we shall consider only Beauville surfaces of unmixed type so that $G_0 = G$.

An extensive research on Beauville surfaces was initiated by the collaboration of Bauer, Catanese and Grunewald [2, 3]. They have been able to intrinsically characterize the groups appearing in minimal presentations of unmixed Beauville surfaces in terms of the existence of the so-called unmixed *Beauville structure*.

Definition 1 An *unmixed Beauville structure* for a finite group G is a quadruple $(x_1, y_1; x_2, y_2)$ of elements of G, which determines two triples (x_1, y_1, z_1) and (x_2, y_2, z_2) satisfying:

(i) $x_1 y_1 z_1 = 1$ and $x_2 y_2 z_2 = 1$,
(ii) $\langle x_1, y_1 \rangle = G$ and $\langle x_2, y_2 \rangle = G$,
(iii) $\Sigma(x_1, y_1, z_1) \cap \Sigma(x_2, y_2, z_2) = \{1\}$,
 where $\Sigma(x, y, z)$ is the union of the conjugacy classes of all powers of x, all powers of y, and all powers of z.

Moreover denoting the order of an element g in G by $|g|$, we define the *type* τ of (x, y, z) to be the triple $(|x|, |y|, |z|)$. In this situation, we say that G admits an *unmixed Beauville structure of type* (τ_1, τ_2).

The question whether a finite group admits an unmixed Beauville structure of a given type is closely related to the question whether it is a quotient of certain triangle groups. More precisely, a necessary condition for a finite group G to admit an unmixed Beauville structure of type $(\tau_1, \tau_2) = \big((r_1, s_1, t_1), (r_2, s_2, t_2)\big)$ is that G is a quotient with torsion free-kernel of the triangle groups T_{r_1,s_1,t_1} and T_{r_2,s_2,t_2}, where

$$T_{r,s,t} = \langle x, y, z : x^r = y^s = z^t = xyz = 1 \rangle.$$

Indeed, conditions (i) and (ii) of Definition 1 are equivalent to the condition that G is a quotient of each of the triangle groups $T_{|x_i|,|y_i|,|z_i|}$, for $i \in \{1, 2\}$, with torsion-free kernel.

When investigating the existence of an unmixed Beauville structure for a finite group, one can consider only types (τ_1, τ_2), where for $i \in \{1, 2\}$, $\tau_i = (r_i, s_i, t_i)$ satisfies $1/r_i + 1/s_i + 1/t_i < 1$. Then T_{r_i,s_i,t_i} is a (infinite non-soluble) hyperbolic triangle group and we say that τ_i is *hyperbolic*. Indeed, if $1/r_i + 1/s_i + 1/t_i > 1$ then T_{r_i,s_i,t_i} is a finite group, and moreover, it is either dihedral or isomorphic to one of A_4, A_5 or S_4. By [2, Proposition 3.6 and Lemma 3.7], in these cases G cannot admit an unmixed Beauville structure. If $1/r_i + 1/s_i + 1/t_i = 1$ then T_{r_i,s_i,t_i} is one of the (soluble infinite) "wall-paper" groups, and by [2, Sect. 6], none of its finite quotients can admit an unmixed Beauville structure.

Observe that condition (iii) of Definition 1 is clearly satisfied under the assumption that $r_1 s_1 t_1$ is coprime to $r_2 s_2 t_2$. However this assumption is not always necessary, as demonstrated by many examples, such as Beauville's original construction, abelian groups [2, Theorem 3.4], alternating groups [20, Theorem 1.2] and the group $\mathrm{PSL}_2(p^e)$ [18].

2 Beauville Surfaces and Finite Simple Groups

A considerable effort has been made to classify the finite simple groups which admit an unmixed Beauville structure. We recall that by *the classification theorem of finite simple groups*, any finite simple group belongs to one of the following families: the cyclic groups Z_p of prime order; the alternating permutation groups $A_n (n \geq 5)$; the finite simple groups of Lie type, defined over finite fields (e.g. $\mathrm{PSL}_n(q)$); and finally the 26 so-called sporadic groups.

A finite abelian simple group clearly does not admit an unmixed Beauville structure as given a prime p, any pair (a, b) of elements of the cyclic group Z_p of prime order p generating it satisfies $\Sigma(a, b, c) = Z_p$. In fact Bauer, Catanese and Grunewald showed in [2, Theorem 3.4] that the only finite abelian groups admitting an unmixed Beauville structure are the abelian groups of the form $Z_n \times Z_n$ where n is a positive integer coprime to 6. (Here Z_n denotes a cyclic group of order n.)

In [2] the authors also provide the first results on finite non-abelian simple groups admitting an unmixed Beauville structure. More precisely they show that the alternating groups of sufficiently large order admit an unmixed Beauville structure, as well as the projective special linear groups $\mathrm{PSL}_2(p)$ where $p > 5$ is a prime. Moreover using computational methods, they checked that every finite non-abelian simple group of order less than 50,000 admits an unmixed Beauville structure with the exception of the alternating group A_5. Based on these results and the latter observation, they conjectured that all finite non-abelian simple groups admit an unmixed Beauville structure with the exception of A_5.

This conjecture has received much attention and has recently been proved to hold. Concerning the simple alternating groups, it was established in [16] that A_5 is indeed the only one not admitting an unmixed Beauville structure. In [17, 20], the conjecture is shown to hold for the projective special linear groups $\mathrm{PSL}_2(q)$ (where $q > 5$), the Suzuki groups $^2B_2(q)$ and the Ree groups $^2G_2(q)$ as well as other families of finite simple groups of Lie type of small rank. More precisely, the projective special and unitary groups $\mathrm{PSL}_3(q)$, $\mathrm{PSU}_3(q)$, the simple groups $G_2(q)$ and the Steinberg triality groups $^3D_4(q)$ are shown to admit an unmixed Beauville structure if q is large (and the characteristic p is greater than 3 for the simple exceptional groups of type G_2 or $^3D_4$). The next major result concerning the investigation of the conjecture with respect to the finite simple groups of Lie type was pursued by Garion, Larsen and Lubotzky who showed in [19] that the conjecture holds for finite non-abelian simple groups of sufficiently large order. The final step regarding the investigation of the conjecture was carried out by Guralnick and Malle [25] and Fairbairn et al. [14] who established its veracity in general, namely,

Theorem 2 ([25]) *Any finite non-abelian simple group, except A_5, admits an unmixed Beauville structure.*

There has also been an effort to classify the finite quasisimple groups and almost simple groups which admit an unmixed Beauville structure. Recall that a finite group G is *quasisimple* provided $G/Z(G)$ is a non-abelian simple group and $G = [G, G]$. In [17] it was shown that $\mathrm{SL}_2(q)$ (for $q > 5$) admits an unmixed Beauville structure. Fairbairn, Magaard and Parker [14] proved the following general result.

Theorem 3 ([14]) *With the exceptions of $\mathrm{SL}_2(5)$ and $\mathrm{PSL}_2(5) \cong \mathrm{SL}_2(4) \cong A_5$, every finite quasisimple group admits an unmixed Beauville structure.*

Recall that a group G is called *almost simple* if there is a non-abelian simple group G_0 such that $G_0 \leq G \leq \mathrm{Aut}\,(G_0)$. By [3, 16] the symmetric groups S_n (where $n \geq 5$) admit an unmixed Beauville structure, and by [18] the group $\mathrm{PGL}_2(p^e)$ admits such a structure.

Moreover for the alternating and symmetric groups Garion and Penegini [20] proved another conjecture that Bauer, Catanese and Grunewald proposed in [2], that almost all of these groups admit a Beauville structure with fixed *type*, namely,

Theorem 4 ([20, Theorem 1.2]) *If $\tau_1 = (r_1, s_1, t_1)$ and $\tau_2 = (r_2, s_2, t_2)$ are two hyperbolic types, then almost all alternating groups A_n admit an unmixed Beauville structure of type (τ_1, τ_2).*

A similar theorem also applies for symmetric groups, see [20], and a similar conjecture was raised in [20], replacing A_n by a finite simple classical group of Lie type of sufficiently large Lie rank, namely,

Conjecture 5 ([20, Conjecture 1.7]) *Let $\tau_1 = (r_1, s_1, t_1)$ and $\tau_2 = (r_2, s_2, t_2)$ be two hyperbolic types. If G is a finite simple classical group of Lie type of Lie rank large enough, then it admits an unmixed Beauville structure of type (τ_1, τ_2).*

In contrast, when the Lie rank is very small, as in the case of $\mathrm{PSL}_2(q)$, such a conjecture does not hold, as demonstrated in [18], where there is a characterization of the types of Beauville structures for these groups.

It is well known that almost all pairs of elements in a finite simple (non-abelian) group are generating pairs [12, 26, 31], hence the following question was raised in the workshop "Beauville surfaces and groups 2012".

Question 6 Let G be a finite (non-abelian) simple group. What is the probability $P(G)$ that for four random elements $x_1, y_1, x_2, y_2 \in G$ the quadruple $(x_1, y_1; x_2, z_2)$ is an unmixed Beauville structure for G?

In particular, is it true that if $G = A_n$ or $G = G_n(q)$, a finite simple group of Lie type of Lie rank n, then $P(G) \to 1$ as $n \to \infty$?

Two interesting comments were made during the workshop regarding this question. The first comment, due to Malle, is that for finite simple groups of Lie type of bounded Lie rank, $P(G)$ does not go to 1, and it is bounded above by a function of the rank. The second comment, due to Magaard, is that the techniques in [14] demonstrate that one can generate many unmixed Beauville structures for the finite simple groups of Lie type, allowing to obtain a constant lower bound on $P(G)$, when G is a finite simple classical group. In the specific case where $G = \mathrm{PSL}_2(q)$ we give the following bounds on the probability of $P(G)$ (see Sect. 4.2 for the proof).

Theorem 7 *Let $G = \mathrm{PSL}_2(q)$.*

- *If q is odd then $\frac{1}{32} - \epsilon_q \le P(G) \le \frac{15}{16} + \epsilon_q$,*
- *If q is even then $\frac{1}{32} - \epsilon_q \le P(G) \le \frac{35}{36} + \epsilon_q$,*

where $\epsilon_q \to 0$ as $q \to \infty$.

3 Hyperbolic Triangle Groups and Their Finite Quotients

Since for a finite group G which admits an unmixed Beauville structure there exists an epimorphism from a hyperbolic triangle group to G, we recall in this section some results on finite quotients of hyperbolic triangle groups.

A *hyperbolic triangle group T* is a group with presentation

$$T = T_{r,s,t} = \langle x, y, z : x^r = y^s = z^t = xyz = 1 \rangle,$$

where (r, s, t) is a triple of positive integers satisfying the condition $1/r + 1/s + 1/t < 1$. Geometrically, let Δ be a hyperbolic triangle group having angles of sizes

$\pi/r, \pi/s, \pi/t$, then T can be viewed as the group generated by rotations of angles $\pi/r, \pi/s, \pi/t$ around the corresponding vertices of Δ in the hyperbolic plane $\mathbb{H}^2$. Moreover, a hyperbolic triangle group $T_{r,s,t}$ has positive measure $\mu(T_{r,s,t})$ where $\mu(T_{r,s,t}) = 1 - (1/r + 1/s + 1/t)$. As hyperbolic triangle groups are infinite and non-soluble it is interesting to study their finite quotients, particularly the simple ones.

A hyperbolic triangle group $T_{r,s,t}$ has minimal measure when $(r, s, t) = (2, 3, 7)$. The group $T_{2,3,7}$ is also called the $(2, 3, 7)$-triangle group and its finite quotients are also known as *Hurwitz groups*. These are named after Hurwitz who showed in the late nineteenth century that if S is a compact Riemann surface of genus $h \geq 2$ then $|\text{Aut } S| \leq 84(h - 1)$ and this bound is attained if and only if Aut S is a quotient of the triangle group $T_{2,3,7}$. Following this result, much effort has been given to classify Hurwitz groups, especially the simple ones, see for example [8] for a historical survey, and [9, 46] for the current state of the art.

Most alternating groups are Hurwitz as shown by Conder (following Higman) who proved in [7] that if $n > 167$ then the alternating group A_n is a quotient of $T_{2,3,7}$. Concerning the finite simple groups of Lie type, there is a dichotomy with respect to their occurrence as quotients of $T_{2,3,7}$ depending on whether the Lie rank is large or not. Indeed as shown in [36] many classical groups of large rank are Hurwitz (and there is no known example of classical groups of large rank which are not Hurwitz). As an illustration by [37] if $n \geq 267$ then the projective special linear group $\text{PSL}_n(q)$ is Hurwitz for any prime power q. The behavior of finite simple groups of Lie type of relatively low rank with respect to the Hurwitz generation problem is rather sporadic. As an illustration by respective results of [6, 38, 39, 46], $\text{PSL}_3(q)$ is Hurwitz if and only if $q = 2$, $\text{PSL}_4(q)$ is never Hurwitz, $G_2(q)$ is Hurwitz for $q \geq 5$, and $\text{PSL}_2(p^e)$ is Hurwitz if and only if $e = 1$ and $p \equiv 0, \pm 1 \bmod 7$, or $e = 3$ and $p \equiv \pm 2, \pm 3 \bmod 7$. Therefore, unlike the alternating groups, there are finite simple groups of Lie type of large order which are not quotients of $T_{2,3,7}$. As for the 26 sporadic finite simple groups, 12 of them are Hurwitz (including the Monster [49]) while the other 14 groups are not.

Turning to general hyperbolic triples (r, s, t) of integers, Higman had already conjectured in the late 1960s that every hyperbolic triangle group has all but finitely many alternating groups as quotients. This was eventually proved by Everitt [13], namely,

Theorem 8 ([13]) *For any hyperbolic triangle group $T = T_{r,s,t}$, if $n \geq n_0(r, s, t)$ then the alternating group A_n is a quotient of T.*

Later, Liebeck and Shalev [32] gave an alternative proof to Higman's Conjecture based on probabilistic group theory, and moreover they have conjectured the following.

Conjecture 9 ([33]) *For any hyperbolic triangle group $T = T_{r,s,t}$, if $G = G_n(q)$ is a finite simple classical group of Lie rank $n \geq n_0(r, s, t)$, then the probability that a randomly chosen homomorphism from T to G is an epimorphism tends to 1 as $|G| \to \infty$.*

This conjecture has been proved by Marion [42, 43] for certain families of groups of small Lie rank and certain triples (r, s, t). For example, take (r, s, t) to be a hyperbolic triple of odd primes and $G = \mathrm{PSL}_3(q)$ or $\mathrm{PSU}_3(q)$ containing elements of orders r, s and t, then the conjecture holds. As another example, if (r, s, t) is a hyperbolic triple of primes and $G = {}^2B_2(q)$ or ${}^2G_2(q)$ contains elements of orders r, s and t, then the conjecture also holds.

However, for finite simple groups of small Lie rank such a conjecture does not hold *in general*, and it fails to hold in the case of $\mathrm{PSL}_2(q)$. Indeed, Langer and Rosenberger [28] and Levin and Rosenberger [30] had generalized the aforementioned result of Macbeath, and determined, for a given prime power $q = p^e$, all the triples (r, s, t) such that $\mathrm{PSL}_2(q)$ is a quotient of $T_{r,s,t}$, with torsion-free kernel. It follows that if (r, s, t) is hyperbolic, then for almost all primes p, there is precisely one group of the form $\mathrm{PSL}_2(p^e)$ or $\mathrm{PGL}_2(p^e)$ which is a homomorphic image of $T_{r,s,t}$ with torsion-free kernel. We note that this result can also be obtained by using other techniques. Firstly, Marion [40] has recently provided a proof for the case where r, s, t are primes relying on probabilistic group theoretical methods. Secondly, it also follows from the representation theoretic arguments of Vincent and Zalesski [48, Theorems 2.9 and 2.11]. Such methods can be used for dealing with other families of finite simple groups of Lie type, see for example [10, 41, 45, 47, 48].

Recently, a new approach was presented by Larsen et al. [29], based on the theory of representation varieties (via deformation theory). They prove a conjecture of Marion [41] showing that various finite simple groups are *not* quotients of $T_{r,s,t}$, as well as positive results showing that many finite simple groups are quotients of $T_{r,s,t}$.

4 Beauville Structures for the Group $\mathrm{PSL}_2(q)$

In this section we discuss the specific case of $\mathrm{PSL}_2(q)$, and briefly sketch the proof of Garion and Penegini [20] for the following theorem, which is based on results of Macbeath [38].

Theorem 10 ([17, 20]). *Let p be a prime number, and assume that $q = p^e$ is at least 7. Then the group $\mathrm{PSL}_2(q)$ admits an unmixed Beauville structure.*

In addition, we bound the probability that four random elements in $\mathrm{PSL}_2(q)$ generate an unmixed Beauville structure and prove Theorem 7.

4.1 Sketch of the Proof of Theorem 10

In order to construct an unmixed Beauville structure for $\mathrm{PSL}_2(q)$ one needs to find a quadruple $(A_1, B_1; A_2, B_2)$ of elements of $\mathrm{PSL}_2(q)$ satisfying the three conditions given in Definition 1. This can be done directly by finding specific elements in the group satisfying these conditions (see [17]) or indirectly by using the following results of Macbeath [38] (see [18, 20]).

Theorem 11 ([38, Theorem 1]) *For every $\alpha, \beta, \gamma \in \mathbb{F}_q$ there exist three matrices $A, B, C \in \mathrm{SL}_2(q)$ satisfying* $\mathrm{tr}(A) = \alpha$, $\mathrm{tr}(B) = \beta$, $\mathrm{tr}(C) = \gamma$ *and* $ABC = I$.

This theorem immediately implies Condition (i).

Moreover, Macbeath [38] classified the pairs of elements in $\mathrm{PSL}_2(q)$ in a way which makes it easy to decide what kind of subgroup they generate. By [38, Theorem 2], a triple $(\alpha, \beta, \gamma) \in \mathbb{F}_q^3$ is *singular*, namely $\alpha^2 + \beta^2 + \gamma^2 - \alpha\beta\gamma - 4 = 0$, if and only if for the corresponding triple of matrices (A, B, C), the group generated by the images of A and B is a *structural subgroup* of $\mathrm{PSL}_2(q)$, that is a subgroup of the Borel or a cyclic subgroup.

Hence, in order to verify Condition (ii), one needs to show that the subgroup generated by $A, B \in \mathrm{PSL}_2(q)$ is neither a structural subgroup (using the aforementioned result of Macbeath), not a dihedral subgroup, not one of the small subgroups A_4, S_4 or A_5, and not a subfield subgroup (namely, isomorphic to $\mathrm{PSL}_2(q_1)$ or to $\mathrm{PGL}_2(q_1)$, where $q = q_1^m$) hence it must be $\mathrm{PSL}_2(q)$ itself, as the subgroup structure of $\mathrm{PSL}_2(q)$ is well-known (see e.g. [11, 44]). For example, Condition (ii) is always satisfied when $q \geq 13$ and the orders $|A| = |B| = |C| = (q-1)/d$ or $(q+1)/d$, where $d = \gcd(2, q-1)$.

Condition (iii) is clearly satisfied under the assumption that the product of the orders $|A_1| \cdot |B_1| \cdot |C_1|$ is coprime to $|A_2| \cdot |B_2| \cdot |C_2|$. For example, for any $q > 7$ the group $\mathrm{PSL}_2(q)$ admits unmixed Beauville structures of types

$$\left((\frac{q-1}{d}, \frac{q-1}{d}, \frac{q-1}{d}), (\frac{q+1}{d}, \frac{q+1}{d}, \frac{q+1}{d}) \right),$$

and

$$\left((\frac{q-1}{d}, \frac{q-1}{d}, \frac{q-1}{d}), (\frac{q+1}{d}, \frac{q+1}{d}, p) \right),$$

appearing in [17, 20] respectively.

However this assumption is not always necessary. Indeed, by [18], $\mathrm{PSL}_2(q)$ (where $q = p^{2e}$, p an odd prime) always admits unmixed Beauville structures of types $((p, p, t_1), (p, p, t_2))$ for certain t_1, t_2 dividing $(q-1)/2$, $(q+1)/2$ respectively.

This approach can be effectively used to construct many unmixed Beauville structures for $\mathrm{PSL}_2(q)$, and in [18] there is a characterization of the types of unmixed Beauville structures for this group.

4.2 Proof of Theorem 7

The proof relies on considering the various types of elements in $G = \mathrm{PSL}_2(q)$. Recall that an element in G is called *split* if its order divides $(q-1)/d$ (where $d = \gcd(2, q-1)$), *non-split* if its order divides $(q+1)/d$, and *unipotent* if its order is p (and then the trace of its pre-image in $\mathrm{SL}_2(q)$ equals ± 2).

It is well-known that there are roughly q^3 matrices in $SL_2(q)$, and moreover, for any $\alpha \in \mathbb{F}_q$ the number of matrices $A \in SL_2(q)$ with $\mathrm{tr}(A) = \alpha$ is roughly q^2 (see for example [1, Table 1]). In addition, the probability that a random element in $\mathbb{F}_q$ is a trace of a split (respectively, non-split) matrix in $SL_2(q)$ goes to $1/2$ as $q \to \infty$ (see [38, Lemma 2]). Therefore, probabilistically, the number of unipotents in G is negligible, and moreover, if we denote by P_q^s (respectively P_q^n) the probability that a random element in G is split (respectively, non-split) then $P_q^s \to \frac{1}{2}$ and $P_q^n \to \frac{1}{2}$ as $q \to \infty$.

By [1, Proposition 7.2] it follows that for any non-singular triple $(\alpha, \beta, \gamma) \in \mathbb{F}_q^3$, the number of triples $(A, B, C) \in SL_2(q)^3$ satisfying $\mathrm{tr}(A) = \alpha, \mathrm{tr}(B) = \beta, \mathrm{tr}(C) = \gamma$ and $ABC = I$ is roughly q^3. By [38, Lemma 3] almost all triples in $\mathbb{F}_q^3$ are non-singular. Since the probability that a random triple of element in $\mathbb{F}_q^3$ contains only traces of split (respectively, non-split) matrices goes to $\frac{1}{8}$ as $q \to \infty$, it follows that the probability that two random elements $A, B \in G$ satisfy that A, B and AB are all split (respectively, non-split), goes to $\frac{1}{8}$ as $q \to \infty$.

In order to obtain a lower bound, observe that the probability that two random elements $A, B \in G$ do **not** generate G goes to 0 as $q \to \infty$ (this can be deduced from the aformentioned results of Macbeath [38], or alternatively, as a specific case of [26]). Namely,

$$\frac{\#\{(A, B) \in G^2 : \langle A, B \rangle \neq G\}}{|G|^2} \le \epsilon_q',$$

where $\epsilon_q' \to 0$ as $q \to \infty$. Hence,

$$\frac{\#\{(A, B) \in G^2 : A, B, AB \text{ are split and } \langle A, B \rangle = G\}}{|G|^2} \ge \frac{1}{8} - \epsilon_q',$$

and

$$\frac{\#\{(A, B) \in G^2 : A, B, AB \text{ are non-split and } \langle A, B \rangle = G\}}{|G|^2} \ge \frac{1}{8} - \epsilon_q'.$$

Since $(q-1)/d$ and $(q+1)/d$ are relatively prime then Condition (iii) is immediately satisfied when A_1, B_1, A_1B_1 are all split and A_2, B_2, A_2B_2 are all non-split (and vice-versa). Therefore,

$$\frac{\#\{(A_1, B_1, A_2, B_2) \in G^4 : (A_1, B_1; A_2, B_2) \text{ is a Beauville structure}\}}{|G|^4}$$

$$\ge \left(\frac{1}{8} - \epsilon_q'\right)^2 + \left(\frac{1}{8} - \epsilon_q'\right)^2 \ge \frac{1}{32} - \epsilon_q,$$

where $\epsilon_q \to 0$ as $q \to \infty$.

Assume that q is odd. In order to obtain an upper bound, observe that if the orders of A_1 and A_2 are both even then Condition (iii) is not satisfied, see [18, Lemma 4.2]. Denote by P_q^e the probability that a random element in G has even order, then $P_q^e \geq \frac{1}{4} - \epsilon'_q$, where $\epsilon'_q \to 0$ as $q \to \infty$. Indeed, if $(q-1)/2$ (respectively, $(q+1)/2$) is even, then at least half of the split (respectively, non-split) elements are of even order, since any split (respectively, non-split) element belongs to a cyclic subgroup of order $(q-1)/2$ (respectively, $(q+1)/2$), and at least half the elements in a cyclic group of even order are of even order. Therefore,

$$\frac{\#\{(A_1, B_1, A_2, B_2) \in G^4 : (A_1, B_1; A_2, B_2) \text{ is } \textbf{not} \text{ a Beauville structure}\}}{|G|^4}$$

$$\geq \left(\frac{1}{4} - \epsilon'_q\right)^2 \geq \frac{1}{16} - \epsilon_q,$$

where $\epsilon_q \to 0$ as $q \to \infty$.

When q is even the proof is similar, replacing P_q^e by the probability that a random element in G has order divisible by 3, which is at least $\frac{1}{6} - \epsilon'_q$, where $\epsilon'_q \to 0$ as $q \to \infty$.

5 Beauville Structures for Finite Simple Groups

In this section we briefly describe the probabilistic group-theoretical methods in proving Theorems 2 and 4. We shall mainly sketch the proof of Garion, Larsen and Lubotzky [19] that the conjecture of Bauer, Catanese and Grunewald (Theorem 2) holds for *almost* all finite simple groups of Lie type, as well as present the proof of Garion and Penegini [20] regarding Beauville structures of alternating groups (Theorem 4), which is based on the probabilistic results of Liebeck and Shalev [32].

As in the probabilistic approach we can ignore finitely many simple groups, we will not deal here with the sporadic groups, whose unmixed Beauville structures can be found in [14, 25]. Thus we shall consider only the alternating groups and the finite simple groups of Lie type.

Recall that in order to construct an unmixed Beauville structure for a finite simple (non-abelian) group G one needs to find a quadruple $(x_1, y_1; x_2, y_2)$ of elements of G satisfying the three conditions given in Definition 1.

5.1 Choosing Disjoint Conjugacy Classes

One usually starts by looking for proper conjugacy classes $X_1, Y_1, Z_1, X_2, Y_2, Z_2$ such that

$$\Sigma(x_1, y_1, z_1) \cap \Sigma(x_2, y_2, z_2) = \{1\}$$

for any $x_i \in X_i$, $y_i \in Y_i$, $z_i \in Z_i$ $(i = 1, 2)$, so that Condition (iii) is satisfied.

For finite simple groups of Lie type, one can choose two maximal tori T_1 and T_2, such that if C_i denotes the set of all conjugates of elements of T_i, then $C_1 \cap C_2 = \{1\}$.

For example, let $G = \mathrm{SL}_{r+1}(q)$ $(r > 1)$, and let t_1 and t_2 denote elements of G whose characteristic polynomials are respectively irreducible (of degree $r + 1$) and the product of irreducible polynomials of degree 1 and r. If T_1 and T_2 denote the centralizers of t_1 and t_2 respectively, then, by [19, Proposition 7], for all $g \in G$, $T_1 \cap g^{-1}T_2g = Z(G)$, thus Condition (iii) is satisfied for $\mathrm{PSL}_{r+1}(q)$.

In fact, for the finite simple groups of Lie type, one can choose several maximal tori T_1 and T_2 such that Condition (iii) is satisfied, see the various choices in [14, 19, 25]. However, the number of conjugacy classes of maximal tori is bounded above by a function of the Lie rank r, and any maximal torus is isomorphic to a product of at most r cyclic groups, so a similar argument to the one presented in Sect. 4.2 implies that the probability that four random elements generate an unmixed Beauville structure is bounded above by a function of r.

One can also choose proper conjugacy classes for the alternating groups (see [14, 17, 20, 25]). In [20], Garion and Penegini used the so-called *almost homogeneous* conjugacy classes introduced by Liebeck and Shalev [32].

Definition 12 ([32]) Conjugacy classes in S_n of cycle-shape (m^k), where $n = mk$, namely, containing k cycles of length m each, are called *homogeneous*. A conjugacy class having cycle-shape $(m^k, 1^f)$, namely, containing k cycles of length m each and f fixed points, with f bounded, is called *almost homogeneous*.

By [20, Algorithm 3.5] one can construct for any six integers $k_1, l_1, m_1, k_2, l_2, m_2 \geq 2$, six distinct almost homogeneous conjugacy classes $X_1, Y_1, Z_1, X_2, Y_2, Z_2$ in A_n whose elements are of orders $k_1, l_1, m_1, k_2, l_2, m_2$ respectively, such that all of them have different numbers of fixed points, thus satisfying Condition (iii).

5.2 Frobenius Formula and Witten's Zeta Function

Condition (i) follows from a classical formula of Frobenius: If X, Y and Z are conjugacy classes in a finite group G, then the number $N_{X,Y,Z}$ of solutions of $xyz = 1$ with $x \in X$, $y \in Y$ and $z \in Z$ is given by

$$N_{X,Y,Z} = \frac{|X| \cdot |Y| \cdot |Z|}{|G|} \sum_{\chi \in \mathrm{Irr}(G)} \frac{\chi(x)\chi(y)\chi(z)}{\chi(1)}, \tag{1}$$

where $\mathrm{Irr}(G)$ denotes the set of complex irreducible characters of G.

Usually, the main contribution to this character sum comes from the trivial character, and the absolute sum on all other characters is negligible. In order to show this, one needs to bound the absolute value $|\chi(g)|$ for an irreducible character χ and an element g, from any of the above conjugacy classes (see Sect. 5.3 for details). If this value can be effectively bounded then one deduces that the value of the sum of the contribution of all *non-trivial* characters to (1) is bounded above by some global constant (depending only on the sizes of G and the conjugacy classes) multiplied by the sum $\sum_{\chi \neq 1} \chi(1)^{-1}$. So it remains to prove that the letter sum converges to 0 as $|G| \to \infty$.

Therefore, a key role in the probabilistic approach is played by the so called *Witten zeta function*, which is defined by

$$\zeta^G(s) = \sum_{\chi \in \mathrm{Irr}(G)} \chi(1)^{-s}.$$

It was originally defined and studied by Witten [50] for Lie groups. For finite simple groups it was studied and applied in detail by Liebeck and Shalev [32–34], who proved the following desired results.

Theorem 13 ([33, Theorem 1.1], [34, Corollary 1.3] and [32, Corollary 2.7]) *Let G be a finite simple group.*

- *If $s > 1$ then $\zeta^G(s) \to 1$ as $|G| \to \infty$.*
- *If $s > 2/3$ and $G \neq \mathrm{PSL}_2(q)$ then $\zeta^G(s) \to 1$ as $|G| \to \infty$.*
- *If $s > 1/2$ and $G \neq \mathrm{PSL}_2(q), \mathrm{PSL}_3(q), \mathrm{PSU}_3(q)$ then $\zeta^G(s) \to 1$ as $|G| \to \infty$.*
- *If $s > 0$ and $G = A_n$ then $\zeta^G(s) \to 1$ as $|G| \to \infty$. Moreover, $\zeta^G(s) = 1 + O(n^{-s})$.*

An alternative approach to prove Condition (i) for the finite simple groups of Lie type, which was successfully applied in [14, 25] and [19, Sect. 4], is based on the following result of Gow [22]: if X and Y are conjugacy classes of regular semisimple elements in a finite Lie type group G, then the set XY contains every non-central semisimple element of G.

5.3 Character Estimates in Finite Simple Groups

A crucial part in the proof is to estimate character values in finite simple groups. More precisely, one needs to bound the absolute value $|\chi(g)|$ for an irreducible character χ and an element g, from any of the conjugacy classes chosen in Sect. 5.1. We therefore recall some useful results for the finite simple groups of Lie type and the alternating groups.

Lemma 14 ([19], Corollary 4) *Let χ be an irreducible character of $G = \mathrm{SL}_{r+1}(q)$.*

- *If $t_1 \in G$ has an irreducible characteristic polynomial, then $|\chi(t_1)| \leq \frac{2(r+1)^2}{\sqrt{3}}$.*
- *If $r \geq 2$ and the characteristic polynomial of $t_2 \in G$ has an irreducible factor of degree r, then $|\chi(t_2)| \leq \frac{2r^2}{\sqrt{3}}$.*

More generally, by [19, Proposition 7], there exist an absolute constant c such that for every sufficiently large group of Lie type G (of Lie rank r), there exist maximal tori T_1 and T_2 as in Sect. 5.1, and for every regular $t \in T_1 \cup T_2$ and every irreducible character χ of G,

$$|\chi(t)| \leq cr^3.$$

Lemma 15 ([32], Proposition 2.12) *Let $\pi \in S_n$ have cycle-shape $(m^k, 1^f)$. Then for any $\chi \in \mathrm{Irr}(S_n)$ we have*

$$|\chi(\pi)| \leq c \cdot (2n)^{\frac{f+1}{2}} \chi(1)^{\frac{1}{m}},$$

where c depends only on m.

It is also interesting to provide an upper bound on the *character ratio* $|\chi(g)/\chi(1)|$, where G is a finite simple group, $g \in G$ and χ is an irreducible character. Gluck and Magaard [21] computed these bounds for the finite classical groups. Such bounds play a crucial role in the proof of a conjecture of Guralnick and Thompson [23], which is related to the inverse Galois problem, namely, which finite groups occur as Galois groups of algebraic number fields (for details see [15]).

5.4 Finding Generating Pairs

In order to prove Condition (ii) one should show that the set of solutions of $xyz = 1$ with $x \in X$, $y \in Y$ and $z \in Z$ contains a generating pair of G, namely, one should avoid pairs (x, y) contained in maximal subgroups of G. Probabilistically, one expects such a result to hold since almost all pairs of elements in a finite simple (non-abelian) group are generating pairs (see [12, 26, 31]).

Namely, one should estimate the sum

$$\sum_{M \in \max G} |\{(x, y, z) \mid x \in X \cap M, y \in Y \cap M, z \in Z \cap M, xyz = 1\}|,$$

where $\max G$ denotes the set of maximal proper subgroups of G. This quantity is bounded above by

$$\sum_{M \in \max G} |M|^2 = |G|^2 \sum_{M \in \max G} \frac{1}{[G : M]^2} \leq \frac{|G|^2}{m(G)^{1/2}} \sum_{M \in \max G} [G : M]^{-3/2},$$

where $m(G)$ is the minimal index of a proper subgroup of G or, equivalently, the minimal degree of a non-trivial permutation representation of G. By estimates of Landazuri and Seitz [27], there exists an absolute constant c such that if G is a finite simple group of Lie type G of Lie rank r then $m(G) \geq cq^r$. Hence, again one should estimate a 'zeta function' encoding the indices of maximal subgroups of finite simple groups of Lie type, which was investigated by Liebeck et al. [35], who proved the following desired result.

Theorem 16 [35] *If G is a finite simple group, and $s > 1$, then*

$$\lim_{|G| \to \infty} \sum_{M \in \max G} [G : M]^{-s} \to 0.$$

Alternatively, in [14, 25] they relied on more delicate results about maximal subgroups in finite simple groups of Lie type containing special elements, called *primitive prime divisors*, of Guralnick et al. [24].

For sufficiently large alternating groups, Conditions (i) and (ii) follow from the following result of Liebeck and Shalev [32]. If (k, l, m) is hyperbolic then the probability that three random elements $x, y, z \in A_n$, with product 1, from almost homogeneous classes X, Y, Z, of orders k, l, m will generate A_n, tends to 1 as $n \to \infty$. Moreover, this probability is $1 - O(n^{-\mu})$, where $\mu = 1 - (1/k + 1/l + 1/m)$.

Acknowledgments The author would like to thank her co-organizers, Ingrid Bauer and Alina Vdovina, and all the participants in the workshop "Beauville surfaces and groups 2012" for their assistance and useful conversations, as well as the University of Newcastle for hosting the workshop. The author is grateful to the referee for pointing out further relevant references.

References

1. T. Bandman, S. Garion, Surjectivity and Equidistribution of the word $x^a y^b$ on PSL $(2, q)$ and SL $(2, q)$. Int. J. Algebra Comput. **22**(2) (2012)
2. I. Bauer, F. Catanese, F. Grunewald, Beauville surfaces without real structures, *Geometric Methods in Algebra and Number Theory*, vol. 235, Progress in Mathematics (Birkhäuser, Boston, 2005), pp. 1–42
3. I. Bauer, F. Catanese, F. Grunewald, Chebycheff and Belyi polynomials, dessins d'enfants, Beauville surfaces and group theory. Mediterr. J. Math. **3**(2), 121–146 (2006)
4. A. Beauville, Surfaces algébriques complexes, Astérisque, **54** Paris, (1978)
5. F. Catanese, Fibred surfaces, varieties isogenous to a product and related moduli spaces. Am. J. Math. **122**, 1–44 (2000)
6. J. Cohen, On non-Hurwitz groups and non-congruence subgroups of the modular group. Glasg. Math. J. **22**, 1–7 (1981)
7. M.D.E. Conder, Generators for alternating and symmetric groups. J. Lond. Math. Soc. **22**, 75–86 (1980)
8. M.D.E. Conder, Hurwitz groups: a brief survey. Bull. Am. Math. Soc. **23**, 359–370 (1990)
9. M.D.E. Conder, An update on Hurwitz groups. Groups, Complex. Cryptol. **2**, 35–40 (2010)
10. L. Di Martino, M.C. Tamburini, A.E. Zalesski, On Hurwitz groups of low rank. Commun. Algebra **28**(11), 5383–5404 (2000)
11. L.E. Dickson, *Linear Groups with an Exposition of the Galois Field Theory* (Teubner, Leipzig, 1901)
12. J.D. Dixon, The probability of generating the symmetric group. Math. Z. **110**, 199–205 (1969)
13. B. Everitt, Alternating quotients of Fuchsian groups. J. Algebra **223**, 457–476 (2000)
14. B. Fairbairn, K. Magaard, C. Parker, Generation of finite simple groups with an application to groups acting on Beauville surfaces, to appear in Proc. Lond. Math. Soc

15. D. Frohardt, K. Magaard, About a conjecture of Guralnick and Thompson, in *Groups, Difference Sets, and the Monster, Proceedings of the Ohio State Conference on Groups and Geometrie*, ed. by K.T. Arasu, et al. (Walter de Gruyter, Berlin, 1996), pp. 43–54

16. Y. Fuertes, G. González-Diez, On Beauville structures on the groups S_n and A_n. Math. Z. **264**, 959–968 (2010)

17. Y. Fuertes, G. Jones, Beauville surfaces and finite groups. J. Algebra **340**, 13–27 (2011)

18. S. Garion, On Beauville structures for PSL (2, q). arXiv:1003.2792

19. S. Garion, M. Larsen, A. Lubotzky, Beauville surfaces and finite simple groups. J. Reine Angew. Math. **666**, 225–243 (2012)

20. S. Garion, M. Penegini, New Beauville surfaces and finite simple groups. Manuscripta Math. **142**, 391–408 (2013)

21. D. Gluck, K. Magaard, Character and fixed point ratios in finite classical groups. Proc. Lond. Math. Soc. **71**, 547–584 (1995)

22. R. Gow, Commutators in finite simple groups of Lie type. Bull. Lond. Math. Soc. **32**(3), 311–315 (2000)

23. R. Guralnick, J. Thompson, Finite groups of genus zero. J. Algebra **131**, 303–341 (1990)

24. R. Guralnick, C. Praeger, T. Penttila, J. Saxl, Linear groups with orders having certain large prime divisors. Proc. Lond. Math. Soc. **78**, 167–214 (1999)

25. R. Guralnick, G. Malle, Simple groups admit Beauville structures. J. Lond. Math. Soc. **85**(3), 694–721 (2012)

26. W.M. Kantor, A. Lubotzky, The probability of generating a finite classical group. Geom. Dedicata **36**, 67–87 (1990)

27. V. Landazuri, G.M. Seitz, On the minimal degrees of projective representations of the finite Chevalley groups. J. Algebra **32**, 418–443 (1974)

28. U. Langer, G. Rosenberger, Erzeugende endlicher projektiver linearer Gruppen. Results Math. **15**(1–2), 119–148 (1989)

29. M. Larsen, A. Lubotzky, C. Marion, Deformation theory and finite simple quotients of triangle groups I. arXiv:1301.2949

30. F. Levin, G. Rosenberger, Generators of finite projective linear groups, II. Results Math. **17**(1–2), 120–127 (1990)

31. M.W. Liebeck, A. Shalev, The probability of generating a finite simple group. Geom. Dedicata **56**, 103–113 (1995)

32. M.W. Liebeck, A. Shalev, Fuchsian groups, coverings of Riemann surfaces, subgroup growth, random quotients and random walks. J. Algebra **276**, 552–601 (2004)

33. M.W. Liebeck, A. Shalev, Fuchsian groups, finite simple groups and representation varieties. Invent. Math. **159**(2), 317–367 (2005)

34. M.W. Liebeck, A. Shalev, Character degrees and random walks in finite groups of Lie type. Proc. Lond. Math. Soc. (3) **90**(1), 61–86 (2005)

35. M.W. Liebeck, B.M.S. Martin, A. Shalev, On conjugacy classes of maximal subgroups of finite simple groups, and a related zeta function. Duke Math. J. **128**(3), 541–557 (2005)

36. A. Lucchini, M.C. Tamburini, Classical groups of large rank as Hurwitz groups. J. Algebra **219**, 531–546 (1999)

37. A. Lucchini, M.C. Tamburini, J.S. Wilson, Hurwitz groups of large rank. J. Lond. Math. Soc. **61**, 81–92 (2000)

38. A.M. Macbeath, Generators of the linear fractional groups, Number Theory. Proc. Sympos. Pure Math. **XII**, Houston, Tex. (1967), Am. Math. Soc., Providence, R.I. (1969), 14–32

39. G. Malle, Hurwitz groups and $G_2(q)$. Can. Math. Bull. **33**, 349–357 (1990)

40. C. Marion, Triangle groups and PSL $_2(q)$. J. Group Theory **12**, 689–708 (2009)

41. C. Marion, On triangle generation of finite groups of Lie type. J. Group Theory **13**, 619–648 (2010)

42. C. Marion, Triangle generation of finite exceptional groups of low rank. J. Algebra **332**, 35–61 (2011)

43. C. Marion, Random and deterministic triangle generation of three-dimensional classical groups I. Commun. Algebra **41**(3), 797–852 (2013)

44. M. Suzuki, *Group Theory I* (Springer, Berlin, 1982)
45. M.C. Tamburini, A.E. Zalesski, Classical groups in dimension 5 which are Hurwitz, Finite groups 2003 (Walter de Gruyter, Berlin, 2004)
46. M.C. Tamburini, M. Vsemirnov, Hurwitz groups and Hurwitz generation, in *Handbook of Algebra*, vol. 4, ed. by M. Hazewinkel (Elsevier, Amsterdam, 2006), pp. 385–426
47. M.C. Tamburini, M. Vsemirnov, Irreducible $(2, 3, 7)$-subgroups of PGL $_n(F)$ for $n \leq 7$. J. Algebra **300**(1), 339–362 (2006)
48. R. Vincent, A.E. Zalesski, Non-Hurwitz classical groups. LMS J. Comput. Math. **10**, 21–82 (2007)
49. R.A. Wilson, The Monster is a Hurwitz group. J. Group Theory **4**(4), 367–374 (2001)
50. E. Witten, On quantum gauge theories in two dimensions. Commun. Math. Phys. **141**, 153–209 (1991)

The Classification of Regular Surfaces Isogenous to a Product of Curves with $\chi(\mathcal{O}_S) = 2$

Christian Gleißner

Abstract A complex surface S is said to be isogenous to a product if S is a quotient $S = (C_1 \times C_2)/G$ where the C_i's are curves of genus at least two and G is a finite group acting freely on $C_1 \times C_2$. In this article we classify all regular surfaces isogenous to a product with $\chi(\mathcal{O}_S) = 2$ under the assumption that the action of G is unmixed i.e. no element of G exchange the factors of the product $C_1 \times C_2$.

1 Introduction

A complex surface S is said to be *isogenous to a product* if S is a quotient

$$S = (C_1 \times C_2)/G,$$

where the C_i's are curves of genus at least two, and G is a finite group acting freely on $C_1 \times C_2$. Due to Catanese [8] there are two possibilities how G can act on the product $C_1 \times C_2$:

- For all $g \in G$ and $(z, w) \in C_1 \times C_2$ we have $g(z, w) = (g(z), g(w))$. In this case the action is called *diagonal*.
- There exists $g \in G$, such that $g(z, w) = (g(w), g(z))$ for all $(z, w) \in C_1 \times C_2$. In this case the curves C_1 and C_2 are isomorphic.

If the action of G is diagonal, we say that S is of *unmixed type*, else we say that S is of *mixed type*.

Since these surfaces were introduced by Catanese [8] there has been produced a considerable amount of literature. In particular the surfaces isogenous to a product with $\chi(\mathcal{O}_S) = 1$ are completely classified: The Bogomolov-Miyaoka-Yau inequality $K_S^2 \leq 9\chi(\mathcal{O}_S)$ together with Debarre's inequality $K_S^2 \geq 2p_g(S)$ gives $0 \leq q(S) = p_g(S) \leq 4$. By Beauville [3] all minimal surfaces S of general type with $p_g(S) =$

C. Gleißner (✉)
Lehrstuhl Mathematik VIII, Universität Bayreuth, Universitätsstraße 30,
95447 Bayreuth, Germany
e-mail: christian.gleissner@uni-bayreuth.de

© Springer International Publishing Switzerland 2015
I. Bauer et al. (eds.), *Beauville Surfaces and Groups*, Springer Proceedings
in Mathematics & Statistics 123, DOI 10.1007/978-3-319-13862-6_6

$q(S) = 4$ are a product of two genus two curves. A minimal surface S of general type with $p_g(S) = q(S) = 3$ is either the symmetric square of a genus three curve or $S = (F_2 \times F_3)/\tau$, where F_g is a curve of genus g and τ is of order two acting on F_2 as an elliptic involution and on F_3 as a fixed point free involution [9, 14, 19]. The classifications of surfaces isogenous to a product in the remaining cases are: the case $p_g = 0$, $q = 0$, due to Bauer et al. [1], $p_g = 1$, $q = 1$, due to Carnovale and Polizzi [7] and $p_g = 2$, $q = 2$, due to Penegini [17].

Our aim is to give a classification in the case $\chi(\mathcal{O}_S) = 2$ under the assumption that S regular and of unmixed type. We want to mention that these surfaces have the invariants $q(S) = 0$ and $p_g(S) = 1$ like $K3$ surfaces and there are recent constructions of $K3$ surfaces with non-symplectic automorphisms as *product quotient surfaces* by Garbagnati and Penegini [10]. Our main result is the following (see also the Table in 5.9):

Theorem 1.1 *There are exactly* 49 *families of regular surfaces isogenous to a product of unmixed type with* $\chi(\mathcal{O}_S) = 2$.

We will now explain how the paper is organized. In Sect. 2 we explain the basics about surfaces isogenous to a product of curves. In Sect. 3 we recall *Riemann's existence theorem* and introduce the necessary tools from group theory and combinatorics. These facts are used to show that there is an entirely group theoretic description of surfaces isogenous to a product. In Sect. 4 we recall the description of the moduli space of surfaces isogenous to a product, due to Catanese. In Sect. 5 we give an algorithm, which we use to classify all regular surfaces isogenous to a product of unmixed type with $\chi(\mathcal{O}_S) = 2$. The computations are performed with the computer algebra system MAGMA [15].[1] In particular the *Database of Small Groups* and the *Database of Perfect Groups* is used. Afterwards we discuss the output of the computation. Finally we show the classification result.

2 Surfaces Isogenous to a Product

In this section we explain some basic facts about surfaces isogenous to a product. We work over the field of complex numbers $\mathbb{C}$ and use the standard notations from the theory of complex surfaces, see for example [4]. The *self-intersection of the canonical class* is denoted by K_S^2, the *topological Euler number* by $e(S) = \sum_{i=1}^{4} (-1)^i h^i(S, \mathbb{C})$, $\kappa(S)$ is the Kodaira dimension. The *holomorphic-Euler-Poincaré-characteristic* is defined as $\chi(\mathcal{O}_S) := 1 - q(S) + p_g(S)$, where $q(S) := h^1(S, \mathcal{O}_S)$ and $p_g(S) := h^2(S, \mathcal{O}_S)$. The basic objects we consider are the following:

Definition 2.1 A surface S is said to be *isogenous to a product* if S is a quotient

$$S = (C_1 \times C_2)/G,$$

---

[1]The source code is available at www.staff.uni-bayreuth.de/~bt300503/.

where the C_i's are smooth projective curves of genus at least two, and $G \leq Aut(C_1 \times C_2)$ is a finite group of automorphisms acting *freely* on $C_1 \times C_2$.

Immediate consequences of this definition are: The surface S is smooth, projective and of general type, i.e. $\kappa(S) = 2$. The canonical class K_S is ample. In particular S is minimal.

The *self-intersection of the canonical class* K_S^2, the *topological Euler number* and the *holomorphic-Euler-Poincaré-characteristic* can be expressed in terms of the genera $g(C_i)$:

Proposition 2.2 ([8, Theorem 3.4]) *Let* $S = (C_1 \times C_2)/G$ *be a surface isogenous to a product, then*

$$K_S^2 = \frac{8(g(C_1) - 1)(g(C_2) - 1)}{|G|}, \quad e(S) = \frac{1}{2}K_S^2 \quad and \quad \chi(\mathcal{O}_S) = \frac{1}{8}K_S^2.$$

In our case $p_g(S) = 1$ and $q(S) = 0$ we get $K_S^2 = 16$ and $e(S) = 8$, moreover we have the useful relation

$$2|G| = (g(C_1) - 1)(g(C_2) - 1). \tag{1}$$

Remark 2.3 For the rest of the paper we consider the *unmixed* case, where the action of G on the product $C_1 \times C_2$ is diagonal, i.e.

$$G = G \cap (Aut(C_1) \times Aut(C_2)).$$

Since we consider unmixed actions only, we obtain two Galois coverings

$$f_1 : C_1 \to C_1/G, \quad f_2 : C_2 \to C_2/G.$$

By [8, Proposition 3.13] one can assume without loss of generality, that G acts faithfully on C_1 and on C_2. We want to relate the invariants $p_g(S)$ and $q(S)$ with the genera of the curves. To do this, we need the following theorem.

Theorem 2.4 *Let X be a smooth projective variety and G be a finite group acting faithfully on X. If $Y = X/G$ is smooth, then*

$$H^0(Y, \Omega_Y^p) \simeq H^0(X, \Omega_X^p)^G.$$

A proof of this result can be found in [12].
By Künneth's formula [13, pp. 103–104]:

$$H^0(C_1 \times C_2, \Omega_{C_1 \times C_2}^1)^G = H^0(C_1, \Omega_{C_1}^1)^G \oplus H^0(C_2, \Omega_{C_2}^1)^G.$$

According to the previous theorem $q(S) = h^0(C_1 \times C_2, \Omega^1_{C_1 \times C_2})^G$. Since $q(S) = 0$ by assumption, we conclude that $g(C_i/G) = 0$ for both $i = 1, 2$. Thus the holomorphic maps f_i from above are Galois coverings of $\mathbb{P}^1_\mathbb{C}$.

3 Group Theory, Riemann Surfaces and Combinatorics

To give a purely group theoretic description of surfaces isogenous to a product, we introduce the required notation from group theory and combinatorics and recall *Riemann's existence theorem*.

Definition 3.1 Let $T = [m_1, \ldots, m_r] \in \mathbb{N}^r$ be an r-tuple. We define

$$\Theta(T) := -2 + \sum_{i=1}^{r} \left(1 - \frac{1}{m_i}\right)$$

and in case $\Theta(T) \neq 0$

$$\alpha(T) := \frac{4}{\Theta(T)}.$$

For $r \geq 3$ we denote by $\mathcal{N}_r$ the set of all r-tuples $[m_1, \ldots, m_r]$ with the following properties:

- $2 \leq m_1 \leq \ldots \leq m_r$
- $\Theta([m_1, \ldots, m_r]) > 0$
- $\alpha([m_1, \ldots, m_r]) \in \mathbb{N}$
- $m_i \big| \alpha([m_1, \ldots, m_r])$ for all $1 \leq i \leq r$.

The union of all $\mathcal{N}_r$ is defined as $\mathcal{N} := \bigcup_{r \geq 3} \mathcal{N}_r$. An element in $\mathcal{N}$ is called a type. A type T contained in $\mathcal{N}_r$ is said to be of length r, we write $l(T) = r$. Moreover for a type $T = [m_1, \ldots, m_r]$ we use the notation $T = [m_1, \ldots, m_r]_{\alpha(T)}$.

For simplicity we write

$$[a_1^{k_1}, \ldots, a_r^{k_r}] := [\underbrace{a_1, \ldots, a_1}_{k_1\text{-times}}, \ldots, \underbrace{a_r, \ldots, a_r}_{k_r\text{-times}}].$$

In the following lemma we give a classification of all types which satisfy the conditions from Definition 3.1 above. This is the starting point of the classification of regular surfaces isogenous to a product with $\chi(\mathcal{O}_S) = 2$ (see also 3.5).

Lemma 3.2 *There are no types of length r if $r = 7$ or $r \geq 9$. The set $\mathcal{N}$ is finite and given by:*

$$
\mathcal{N} = \left\{
\begin{array}{lllll}
[2,3,7]_{168}, & [2,3,8]_{96}, & [2,4,5]_{80}, & [2,3,9]_{72}, & [2,3,10]_{60}, \\
[2,3,12]_{48}, & [2,4,6]_{48}, & [3^2,4]_{48}, & [2,3,14]_{42}, & [2,5^2]_{40}, \\
[2,3,18]_{36}, & [2,4,8]_{32}, & [2,3,30]_{30}, & [2,5,6]_{30}, & [3^2,5]_{30}, \\
[2,4,12]_{24}, & [2,6^2]_{24}, & [3,4^2]_{24}, & [3^2,6]_{24}, & [2^3,3]_{24}, \\
[3^2,7]_{21}, & [2,4,20]_{20}, & [2,5,10]_{20}, & [2,6,9]_{18}, & [3^2,9]_{18}, \\
[2,8^2]_{16}, & [4^3]_{16}, & [2^3,4]_{16}, & [3,5^2]_{15}, & [3^2,15]_{15}, \\
[2,7,14]_{14}, & [2,12^2]_{12}, & [3,4,12]_{12}, & [3,6^2]_{12}, & [4^2,6]_{12}, \\
[2^3,6]_{12}, & [2^2,3^2]_{12}, & [5^3]_{10}, & [2^3,10]_{10}, & [3,9^2]_{9}, \\
[4,8^2]_{8}, & [2^2,4^2]_{8}, & [2^5]_{8}, & [7^3]_{7}, & [2^2,6^2]_{6}, \\
[2,3^2,6]_{6}, & [3^4]_{6}, & [2^4,3]_{6}, & [4^4]_{4}, & [2^3,4^2]_{4}, \\
[2^6]_{4}, & [3^5]_{3}, & [2^8]_{2} & &
\end{array}
\right.
$$

Proof We use the fourth property in the case $i = r$:

From $m_r \left| \dfrac{4}{-2 + \sum_{i=1}^{r}\left(1 - \frac{1}{m_i}\right)} \right.$ it follows $\displaystyle\sum_{i=1}^{r}\left(1 - \frac{1}{m_i}\right) \leq 2 + \frac{4}{m_r}$ (∗).

Since $m_i \geq 2$ for all $1 \leq i \leq r$, we get

$$
r - 2 \leq \sum_{i=1}^{r}\frac{1}{m_i} + \frac{4}{m_r} \leq \frac{r}{2} + 2,
$$

and therefore $r \leq 8$. We now investigate two cases: $r = 3$ and $r \geq 4$.

- If $r = 3$, we claim that $m_2 \geq 3$. Suppose $m_2 = 2$, then also $m_1 = 2$ and

$$
0 < \Theta([m_1, m_2, m_3]) = 1 - \frac{1}{m_1} - \frac{1}{m_2} - \frac{1}{m_3} = -\frac{1}{m_3},
$$

a contradiction. The inequality (∗) in the case $r = 3$ reads

$$
\frac{5}{m_3} \geq 1 - \frac{1}{m_1} - \frac{1}{m_2} \geq 1 - \frac{1}{2} - \frac{1}{3} = \frac{1}{6}.
$$

From this we conclude $m_3 \leq 30$.

- In the second case $r \geq 4$, we use the formula (∗) again:

$$
r - \sum_{i=1}^{r}\frac{1}{m_i} \leq 2 + \frac{4}{m_r}
$$

and get

$$\frac{5}{m_r} \geq r - 2 - \sum_{i=1}^{r-1} \frac{1}{m_i} \geq \frac{r-3}{2}.$$

Because of this, we always have $10 \geq \frac{10}{r-3} \geq m_r$ in that case.

Now, it suffices to check only a finite number of types. This can be easily done with a computer.

Definition 3.3 Let G be a finite group, $2 \leq m_1 \leq \ldots \leq m_r$ integers. A *spherical system* of generators of G of type $[m_1, \ldots, m_r]$ is an r-tuple $A = (g_1, \ldots, g_r)$ of elements of G, such that:

- $G = \langle g_1, \ldots, g_r \rangle, \quad g_1 \cdot \ldots \cdot g_r = 1_G.$
- There exists a permutation $\tau \in \mathfrak{S}_r$, such that $ord(g_i) = m_{\tau(i)}$.

The *stabilizer set* of A is defined as

$$\Sigma(A) := \bigcup_{h \in G} \bigcup_{j \in \mathbb{Z}} \bigcup_{i=1}^{r} \{h g_i^j h^{-1}\}.$$

A pair (A_1, A_2) of spherical systems of generators of G is called *disjoint*, if and only if

$$\Sigma(A_1) \cap \Sigma(A_2) = \{1_G\}.$$

The geometry behind this definition is known as *Riemann's existence theorem*. A detailed explanation can be found in [16, chapter 3, Sects. 3 and 4]. We will use the following version of this theorem:

Theorem 3.4 (Riemann's existence theorem) *A finite group G acts as a group of automorphisms on a compact Riemann surface C of genus $g(C) \geq 2$, such that $C/G \simeq \mathbb{P}^1_{\mathbb{C}}$ if and only if there exists a spherical system of generators A of G of type $T = [m_1, \ldots, m_r]$, such that the following Riemann-Hurwitz formula holds:*

$$2g(C) - 2 = |G|\Theta(T).$$

By Riemann's existence theorem we have a *group theoretical description* of surfaces isogenous to a product: Given $S = (C_1 \times C_2)/G$, isogenous to a product, we can attach a disjoint pair of spherical systems of generators

$$(A_1(S), A_2(S)) \text{ of type } (T_1(S), T_2(S)).$$

Geometrically, disjoint means that G acts without fixed points on $C_1 \times C_2$. Conversely, the data above determine a surface isogenous to a product.

Next, we want to show that the types $(T_1(S), T_2(S))$, attached to a regular surface S isogenous to a product with $p_g(S) = 1$ of unmixed type, satisfy the conditions of 3.1. The proof of this fact is similar to the proof given in [1]. For convenience of the reader we will present the proof.

Theorem 3.5 *Let S be a surface isogenous to a product of curves of unmixed type with $p_g(S) = 1$ and $q(S) = 0$. Let $T_1(S) = [m_1, \ldots, m_r]$ and $T_2(S) = [n_1, \ldots, n_s]$ be the corresponding types, then*

- $\Theta(T_i(S)) > 0$ *for* $i = 1, 2$.
- $\alpha(T_i(S)) \in \mathbb{N}$ *for* $i = 1, 2$.
- $m_i \big| \alpha(T_1(S))$ *for all* $1 \le i \le r$ *and* $n_i \big| \alpha(T_2(S))$ *for all* $1 \le i \le s$.

Proof We consider the holomorphic maps $f_i : C_i \to C_i/G$ and apply the Riemann-Hurwitz formula

$$2g(C_i) - 2 = |G|\Theta(T_i(S)), \quad i = 1, 2. \tag{2}$$

This already shows the first claim, since $g(C_1) \ge 2$ and $g(C_2) \ge 2$. From (2) and $2|G| = (g(C_1) - 1)(g(C_2) - 1)$ (1), we deduce

$$\alpha(T_1(S)) = \frac{4}{\Theta(T_1(S))} = g(C_2) - 1 \text{ and } \alpha(T_2(S)) = \frac{4}{\Theta(T_2(S))} = g(C_1) - 1$$

so the second claim follows. It remains to prove the third claim. Let $A_1(S) = (g_1, \ldots, g_r)$ be a corresponding ordered spherical system of generators of G of type $T_1(S)$. The cyclic group $\langle g_i \rangle$ of order m_i acts on C_1 with at least one fixed point, but the action on the product $C_1 \times C_2$ is free. Therefore $\langle g_i \rangle$ acts on C_2 freely. The map $C_2 \to C$ of degree m_i, where $C := C_2/\langle g_i \rangle$ is unramified. In this case we have $2g(C) - 2 = \frac{2g(C_2)-2}{m_i} > 0$, due to Riemann-Hurwitz. Hence

$$g(C_2) - 1 = \alpha(T_1(S)) = m_i(g(C) - 1),$$

for all $i = 1, \ldots, r$. With the same argument we can show that $n_i \big| \alpha(T_2(S))$ for all $i = 1, \ldots, s$.

4 Moduli Spaces

In this section we want to describe the moduli space of surfaces isogenous to a product. We follow the papers [1, S31, S8–9] and [18, Appendix]. Due to the work of Gieseker [11] there exists a quasi-projective moduli space of minimal smooth projective surfaces of general type with fixed invariants K_S^2 and $\chi(\mathcal{O}_S)$, which is denoted by $\mathfrak{M}_{(\chi(\mathcal{O}_S), K_S^2)}$. For a fixed finite group G and a fixed pair of types (T_1, T_2) we denote the subset of $\mathfrak{M}_{(2,16)}$ of isomorphism classes of surfaces isogenous to a product, which admit a disjoint pair of spherical systems of generators (A_1, A_2) of type (T_1, T_2), by $\mathfrak{M}_{(G,T_1,T_2)}$.

Theorem 4.1 ([1, Remark 5.1])

- *The subset $\mathfrak{M}_{(G,T_1,T_2)} \subset \mathfrak{M}_{(2,16)}$ consists of a finite number of connected components of the same dimension, which are irreducible in the Zariski topology.*
- *The dimension $d(G, T_1, T_2)$ of any component in $\mathfrak{M}_{(G,T_1,T_2)}$ is*

$$d(G, T_1, T_2) = l(T_1) - 3 + l(T_2) - 3.$$

The problem to determine the number n of the connected components of $\mathfrak{M}_{(G,T_1,T_2)}$ can be translated in a group theoretical problem. We recall the following definition:

Definition 4.2 Let $r \in \mathbb{N}$ be a positive integer. We define the *Artin-Braid group* $\boldsymbol{B}_r$ as

$$\boldsymbol{B}_r := \left\langle \sigma_1, \ldots, \sigma_{r-1} \left| \begin{array}{c} \sigma_i \sigma_j = \sigma_j \sigma_i \quad if \quad |i - j| > 1 \\ \sigma_i \sigma_{i+1} \sigma_i = \sigma_{i+1} \sigma_i \sigma_{i+1} \quad for \quad i = 1, \ldots, r - 2 \end{array} \right. \right\rangle.$$

Let G be a finite group and T be a type of length $l(T) = r$. We denote the set of spherical systems of generators of G of type T by $\mathcal{B}(G, T)$. The Artin-Braid group $\boldsymbol{B}_r$ acts on $\mathcal{B}(G, T)$ as follows:

$$\sigma_i(A) := (g_1, \ldots, g_{i-1}, g_i \cdot g_{i+1} \cdot g_i^{-1}, g_i, g_{i+2}, \ldots, g_r),$$

for all $A = (g_1, \ldots, g_r) \in \mathcal{B}(G, T)$ and $1 \leq i \leq r - 1$. This determines a well-defined action, which is called the *Hurwitz action*. There is also a natural action of $Aut(G)$ on $\mathcal{B}(G, T)$:

$$\varphi(A) := (\varphi(g_1), \ldots, \varphi(g_r)),$$

for all $\varphi \in Aut(G)$. Let $(\gamma_1, \gamma_2, \varphi) \in \boldsymbol{B}_r \times \boldsymbol{B}_s \times Aut(G)$ be a triple and $(A_1, A_2) \in \mathcal{B}(G, T_1) \times \mathcal{B}(G, T_2)$, we define

$$(\gamma_1, \gamma_2, \varphi) \cdot (A_1, A_2) := (\varphi(\gamma_1(A_1)), \varphi(\gamma_2(A_2))).$$

It is easy to verify that this defines an action of $\boldsymbol{B}_r \times \boldsymbol{B}_s \times Aut(G)$ on $\mathcal{B}(G, T_1) \times \mathcal{B}(G, T_2)$. We denote this action by $\mathfrak{T}$, and its restriction to the first factor $\mathcal{B}(G, T_1)$ by $\mathfrak{T}_1$ and to the second factor $\mathcal{B}(G, T_2)$ by $\mathfrak{T}_2$.[2]

Theorem 4.3 ([1, Proposition 5.2]) *Let S and S' be surfaces isogenous to a product of unmixed type with $q(S) = q(S') = 0$. The surfaces S and S' are in the same irreducible component if and only if*

- $G(S) \simeq G(S')$,
- $(T_1(S), T_2(S)) = (T_1(S'), T_2(S'))$,
- *$(A_1(S), A_2(S))$ and $(A_1(S'), A_2(S'))$ are in the same $\mathfrak{T}$-orbit, or $(A_1(S), A_2(S))$ and $(A_2(S'), A_1(S'))$ are in the same $\mathfrak{T}$-orbit.*

---

[2]We want to stress that the action of $Aut(G)$ and the Hurwitz action commute.

In theory we now have a method to compute the number of connected components of $\mathfrak{M}_{(G;T_1,T_2)}$: first we compute a representative (A_1, A_2) for each orbit of the action $\mathfrak{T}$. Then we determine the pairs, where the intersection $\Sigma(A_1) \cap \Sigma(A_2)$ is trivial. If $T_1(S) = T_2(S)$ and there is more than one pair, we have to consider the $\mathbb{Z}_2$ action corresponding to the exchange of the curves.[3] The number of the remaining pairs is the number of connected components of $\mathfrak{M}_{(G;T_1,T_2)}$. However, the set $\mathcal{B}(G, T_1) \times \mathcal{B}(G, T_2)$ can be very large. Even with a computer it is not possible, or at least very time consuming, to perform this calculation. To improve the speed of the calculation we use an idea of Penegini and Rollenske, which is based on the following lemma:

Lemma 4.4 ([18, Appendix 6.1]) *Let $(A_1, A_2), (B_1, B_2) \in \mathcal{B}(G, T_1) \times \mathcal{B}(G, T_2)$.*

- *If A_i and B_i are in the same orbit of the Hurwitz action, then also the pairs (A_1, A_2) and (B_1, B_2) are in the same $\mathfrak{T}$-orbit.*
- *If A_1 and B_1 are in different $\mathfrak{T}_1$-orbits, then also (A_1, A_2) and (B_1, B_2) are in different $\mathfrak{T}$ orbits.*

Now we have an effective algorithm:

- Compute a set $\mathfrak{R}_1$ of representatives of the action $\mathfrak{T}_1$ on $\mathcal{B}(G, T_1)$.
- Compute a set $\mathfrak{R}_2$ of representatives of the Hurwitz action on $\mathcal{B}(G, T_2)$.
- Determine the set of tuples $(A_1, A_2) \in \mathfrak{R}_1 \times \mathfrak{R}_2$ which satisfy:

$$\Sigma(A_1) \cap \Sigma(A_2) = \{1_G\}.$$

This set is denoted by $\mathfrak{R}$.

We achieve the following: Every orbit of $\mathfrak{T}$ has at least one representative in $\mathfrak{R}$ by 4.4. Hence we have an upper bound for the number n of $\mathfrak{T}$-orbits. We also have a lower bound for n. The pairs (A_1, A_2) and (B_1, B_2) are in different orbits of $\mathfrak{T}$, if

1. $A_1 \neq B_1$ or if
2. A_2 and B_2 are in different orbits of $\mathfrak{T}_2$ (cf. 4.4).

- In most cases it is possible to determine the number n of orbits of $\mathfrak{T}$ using this method. If this is not possible, then we exchange the pairs T_1 and T_2 and compute the set $\mathfrak{R}$ again. If it is still not possible to determine n we have to identify the pairs of the smaller set using the action $\mathfrak{T}$.

5 The Algorithm and the Classification Result

In this section we explain our algorithm, which allows us to classify all regular surfaces isogenous to a product of curves of unmixed type with $\chi(\mathcal{O}(S)) = 2$. For the implementation of the algorithm the computer algebra system MAGMA [15]

---

[3]This happens in only one of our examples. Thus we consider only the $\mathfrak{T}$ action in our program and treat the exceptional example separately 5.8.

is used. The program is based on the program in the appendix of [2]. After the explanation of the algorithm, we discuss the output of the computations. Finally we give our classification result.

Let $S = (C_1 \times C_2)/G$ be a regular surface isogenous to a product of unmixed type with $\chi(\mathcal{O}_S) = 2$. From $2|G| = (g(C_1) - 1)(g(C_2) - 1)$, $\alpha(T_1(S)) = g(C_2) - 1$ and $\alpha(T_2(S)) = g(C_1) - 1$ for the attached types $T_i(S)$, it follows

$$|G| = \frac{1}{2}\alpha(T_1)\alpha(T_2).$$

According to the list in Lemma 3.2, $\alpha(T_i) \leq 168$ and therefore $|G| \leq 14112$. The group order is also bounded in terms of the genera $g(C_i)$, in fact $|Aut(C_i)| \leq 84(g(C_i) - 1)$ due to Hurwitz' famous theorem. For small $g(C_i)$ there are better bounds. In Breuer's book [5, p. 91] there is a table which gives the maximum order of $|Aut(C_i)|$ in case $2 \leq g(C_i) \leq 48$.

Definition 5.1 Let $A = [m_1, \ldots, m_r]$ be an r-tuple of integers $m_i \geq 2$. The *polygonal group* $\mathbb{T}(m_1, \ldots, m_r)$ is defined as

$$\mathbb{T}(m_1, \ldots, m_r) = <t_1, \ldots, t_r \quad | \quad t_1 \cdot \ldots \cdot t_r = t_1^{m_1} = \ldots = t_r^{m_r} = 1>.$$

A group G admitting a spherical system of generators of type $[m_1, \ldots, m_r]$ is a quotient of $\mathbb{T}(m_1, \ldots, m_r)$. The following lemma will be used in the sequel. The proof of it is elementary and will be omitted.

Lemma 5.2 *Let G be a group and H a quotient of G, then:*

- *H^{ab} is a quotient of G^{ab}.*
- *The commutator subgroup $[H, H]$ is a quotient of $[G, G]$.*
- *If G is a quotient of $\mathbb{T}(2, 3, 7)$, then G is perfect.*

We can now describe the algorithm briefly. We perform the following steps:

Step 1: The program computes the set $\mathcal{N}$ of types given in Lemma 3.2. For every integer $m \leq 14112$ we compute the set of all triples of the form (m, T_1, T_2) up to permutation of T_1 and T_2, where $T_1, T_2 \in \mathcal{N}$ and $m = \frac{1}{2}\alpha(T_1)\alpha(T_2)$.

Step 2: For every triple (m, T_1, T_2) the script computes $g(C_2) = \alpha(T_1(S)) + 1$ and $g(C_1) = \alpha(T_2(S)) + 1$. If $2 \leq g(C_i) \leq 48$ for at least one i, we check if m is less or equal to the maximum group order of the automorphism group $Aut(C_i)$ allowed by Breuer's table.

Step 3: For every triple (m, T_1, T_2) passing this test, the script searches the list of groups of order m for a group admitting a spherical system of generators of type T_1 and one of type T_2.

Step 4: For each triple (m, T_1, T_2) and each group G of order m, admitting a spherical system of generators of type T_1 and of type T_2, the script computes the number of orbits of the action $\mathfrak{T}$ on $\mathcal{B}(G, T_1) \times \mathcal{B}(G, T_2)$, using the method explained after Lemma 4.4.

In **Step 3** we face two computational difficulties:

- In MAGMA's *Database of Small Groups* all groups of order $|G| \leq 2000$ are contained, except the groups of order $|G| = 1024$ (which can not occur). In the case $2001 \leq |G| \leq 14112$, there is no MAGMA *Database* containing all groups of these orders. Only the perfect groups with $|G| \leq 50000$ are contained in MAGMA's *Database of Perfect Groups*.
- Groups of order

$$|G| \in \{1920, 1152, 768, 512, 384, 256\}$$

can occur. Despite the fact, that we have access to all groups of these orders, it is not efficient to search through all of them for spherical systems of generators, because the number of these groups is too high.[4]

Due to these difficulties we split the program into two *main routines* namely *Mainloop*1 and *Mainloop*2.

- The function *Mainloop*1 treats the cases where

$$|G| \geq 2001 \quad \text{or} \quad |G| \in \{1920, 1152, 768, 512, 384, 256\}.$$

(i) The case $m = |G| \geq 2001$. For all triples (m, T_1, T_2), we search the *Database of Perfect Groups* for a perfect group of order m, admitting a spherical system of generators of type T_1 and T_2. If neither $T_1 = [2, 3, 7]$ nor $T_2 = [2, 3, 7]$ we can not yet decide if there is a non-perfect group of order m, admitting a spherical system of generators of type T_i (cf. 5.2). The script then saves the triple in the file *exceptional.txt*. We have to investigate these cases with theoretical arguments (see Sect. 5.1), and we will show these do not occur.

(ii) The case $|G| \in \{1920, 1152, 768, 512, 384, 256\}$. Each group G, admitting a spherical system of generators of type $T_1 = [n_1, \ldots, n_r]$ and of type $T_2 = [m_1, \ldots, m_s]$ is a quotient of

$$\mathbb{T}(T_1) := \mathbb{T}(n_1, \ldots, n_r) \quad \text{and} \quad \mathbb{T}(T_2) := \mathbb{T}(m_1, \ldots, m_s).$$

According to Lemma 5.2 the group G^{ab} is a quotient of

$$\mathbb{T}(n_1, \ldots, n_r)^{ab} \quad \text{and} \quad \mathbb{T}(m_1, \ldots, m_s)^{ab}.$$

The script first computes all possible abelianizations of G from the types. With this step we can exclude the groups which don't have the right abelianization. Then we search through the remaining groups if there is a group, admitting spherical systems of generators of type T_i.

---

[4]Indeed there are 12.059.590 groups G, such that $|G| \in \{1920, 1152, 768, 512, 384, 256\}$.

- The function $Mainloop2$ treats the case where

$$|G| \leq 2000 \quad \text{and} \quad |G| \notin \{1920, 1152, 768, 512, 384, 256\}.$$

The output is written in the file $loop2.txt$. Due to the high use of memory, if one of the types is $[2^8]$, we split the computation into two parts:

(i) With the command $Mainloop2(n_1, n_2, 1)$ the script classifies, in the sense of above, all surfaces where $n_1 \leq |G| \leq n_2$ and $T_i \neq [2^8]$.
(ii) With the command $Mainloop2(n_1, n_2, 0)$ the script classifies all surfaces where $n_1 \leq |G| \leq n_2$ and $T_1 = [2^8]$.

Also in this main routine, if $T_1 = [2^8]$ we have to treat some cases separately. Our workstation has not enough memory to compute the set $\mathcal{B}(G, [2^8])$, if

$$|G| \in \{168, 96, 48\}.$$

The occurring triples (n, T_1, T_2), where n is one of the group orders above are marked by the script as exceptional. We treat these cases in Sect. 5.2.

Before we discuss the exceptional cases of the output, we explain some notation from group theory, that will be used in this and in the next section.

- We use the following MAGMA notation: $\langle a, b \rangle$ denotes the group of order a having number b in the database of Small Groups [15].
- The group $U(4, 2) \leq Gl(4, \mathbb{F}_2)$ is defined to be the subgroup of upper triangle matrices $A = (a_{ij})$ with $a_{ii} = 1$, for all $1 \leq i \leq 4$.
- The group $G(128, 36) := \langle 128, 36 \rangle$ is given in a polycyclic presentation:

$$\langle 128, 36 \rangle = \left\langle g_1, \ldots, g_7 \left| \begin{array}{lll} g_1^2 = g_4, & g_2^2 = g_5, & g_2^{g_1} = g_2 g_3 \\ g_3^{g_1} = g_3 g_6, & g_3^{g_2} = g_3 g_7, & g_4^{g_2} = g_4 g_6 \\ g_5^{g_1} = g_5 g_7 \end{array} \right. \right\rangle$$

Here $g_i^{g_j}$ means $g_j^{-1} g_i g_j$. The squares of the generators $g_1, \ldots, g_7$, which are not mentioned in the presentation are equal to 1. If $g_i^{g_j} = g_i$, this relation is omitted in the presentation.

5.1 Exceptional Cases for Mainloop1:

The following cases are skipped by the program, because the group order is greater than 2000, and saved in the file $exceptional.txt$:

$|G|$	T_1	T_2
4608	[2, 3, 8]	[2, 3, 8]
3840	[2, 3, 8]	[2, 4, 5]
3456	[2, 3, 8]	[2, 3, 9]
3200	[2, 4, 5]	[2, 4, 5]
2880	[2, 3, 9]	[2, 4, 5]
2880	[2, 3, 8]	[2, 3, 10]
2592	[2, 3, 9]	[2, 3, 9]
2400	[2, 3, 10]	[2, 4, 5]
2304	[2, 3, 8]	[2, 4, 6]
2304	[2, 3, 8]	[3, 3, 4]
2304	[2, 3, 8]	[2, 3, 12]
2160	[2, 3, 9]	[2, 3, 10]

Our aim is to show, that the cases above can not occur.

Proposition 5.3 *There are no finite groups G admitting a disjoint pair of spherical systems of generators (A_1, A_2) of type (T_1, T_2), where $|G|$, T_1 and T_2 are in the table above.*

Proof Our MAGMA code has already excluded the cases where G is perfect. We only treat the case $|G| = 4608$, $T_1 = [2, 3, 8]$ and $T_2 = [2, 3, 8]$. The other cases can be excluded using similar methods. The ideas we use are from [1]. The group G is a quotient of $\mathbb{T}(2, 3, 8)$ and there is a surjective homomorphism $\phi^{ab} : \mathbb{T}(2, 3, 8)^{ab} \to G^{ab}$. Since G is not perfect, $G^{ab} \neq \{1_G\}$. Similarly, the commutator subgroup $G' = [G, G]$ is also a quotient of $\mathbb{T}(2, 3, 8)' = [\mathbb{T}(2, 3, 8), \mathbb{T}(2, 3, 8)]$. We have

$$\mathbb{T}(2, 3, 8)^{ab} \simeq \mathbb{Z}_2, \quad \text{and} \quad \mathbb{T}(2, 3, 8)' \simeq \mathbb{T}(3, 3, 4).$$

This implies $G^{ab} \simeq \mathbb{Z}_2$, thus $|G'| = 2304$ and the group G' is a quotient of $\mathbb{T}(3, 3, 4)$.

$$\mathbb{T}(3, 3, 4)^{ab} \simeq \mathbb{Z}_3, \quad \text{and} \quad \mathbb{T}(3, 3, 4)' \simeq \mathbb{T}(4, 4, 4).$$

Since $|G'| = 2^8 \cdot 3^2$ the group G' is solvable, due to Burnsides theorem [6]. We have $G'^{ab} \simeq \mathbb{Z}_3$, thus: $|G''| = 768$ and G'' is a quotient of $\mathbb{T}(4, 4, 4)$. This is impossible according to the following computational fact, which can be verified with MAGMA.

Lemma 5.4 ([1, Lemma 4.11]) *There are 1090235 groups of order 768. None of them is a quotient of $\mathbb{T}(4, 4, 4)$.*

5.2 Exceptional Cases for Mainloop2:

Here $T_1 = [2^8]$ and $T_2 \in \mathcal{N}$. We have $|G| = \frac{1}{2}\alpha(T_1)\alpha(T_2) = \alpha(T_2) \leq 168$. Since our workstation has not enough memory to compute the set $\mathcal{B}(G, [2^8])$ if $|G| \in \{168, 96, 48\}$, these cases are marked as exceptional. We receive the following output:

T_2	G	No. of $\mathfrak{T}$-orbits	
$[2, 3, 7]$	$\langle 168, 42\rangle$	0	Exceptional case
$[2, 3, 8]$	$\langle 96, 64\rangle$	0	Exceptional case
$[2, 4, 6]$	$\langle 48, 48\rangle$	2	Exceptional case
$[3, 4^2]$	$\langle 24, 12\rangle$	1	
$[2^3, 4]$	$\langle 16, 11\rangle$	2	
$[2^5]$	$\langle 8, 5\rangle$	1	

Next, we will investigate the exceptional cases.

Proposition 5.5 *The group $\langle 168, 42\rangle$ has no disjoint pair of spherical systems of generators of type $([2^8], [2, 3, 7])$.*

Proof A MAGMA computation shows, that this group has only one conjugacy class of elements of order 2. Hence, for every pair (A_1, A_2) of generators of type $([2^8], [2, 3, 7])$ the intersection $\Sigma(A_1) \cap \Sigma(A_2)$ is nontrivial.

Proposition 5.6 *The group $\langle 96, 64\rangle$ has no disjoint pair of spherical systems of generators of type $([2^8], [2, 3, 8])$.*

Proof A MAGMA calculation shows, that this group has two conjugacy classes of elements of order 2. We denote them by K_1 and K_2. We have $|K_2| = 3$ and $\langle K_2\rangle \simeq \mathbb{Z}_2 \times \mathbb{Z}_2$. Since $|\langle K_2\rangle| = 4$, there is no spherical system of generators $A_1 = (h_1, \ldots, h_8)$ of G of type $[2^8]$, with $h_i \in K_2$ for all $1 \leq i \leq 8$. Every spherical system of generators of G contains elements of K_1. A further MAGMA calculation shows, that there is no spherical system of generators $A_2 = (g_1, g_2, g_3)$ of G of type $[2, 3, 8]$, with $g_1 \in K_2$. $\qquad\square$

Proposition 5.7 *The group $\langle 48, 48\rangle$ admits disjoint pairs of spherical systems of generators of type $([2^8], [2, 4, 6])$. The number of $\mathfrak{T}$-orbits is two.*

Proof The group G has 19 elements of order 2, they are contained in 5 conjugacy classes. We denote them by $K_1, \ldots, K_5$ and the set of elements of order two by M.

Class	Rep	Length
K_1	$G.2$	1
K_2	$G.4$	3
K_3	$G.2 * G.4$	3
K_4	$G.1$	6
K_5	$G.1 * G.2$	6

A MAGMA calculation shows, that $g_1 \in K_4 \cup K_5$ for each spherical system of generators $(g_1, g_2, g_3) \in \mathcal{B}(G, [2, 4, 6])$. Since we are interested in disjoint pairs of spherical systems of generators, it is not necessary to compute the whole set $\mathcal{B}(G, [2^8])$. All elements $(h_1, \ldots, h_8) \in \mathcal{B}(G, [2^8])$, which contain some $h_i \in K_4$, as well as some $h_j \in K_5$ are irrelevant. For each $1 \leq l \leq 5$ we define subsets

$$\mathcal{B}_l := \{[h_1, \ldots, h_8] \in \mathcal{B}(G, [2^8]) \mid h_i \in M \setminus K_l\}$$

of $\mathcal{B}(G, [2^8])$ and compute them in the cases $l = 4$ and $l = 5$. We have $|\mathcal{B}_4| = |\mathcal{B}_5| = 9.213.120$. Since the Hurwitz action acts via conjugation and permutation of elements, we can restrict it to $\mathcal{B}_4$ and to $\mathcal{B}_5$. Next we compute for each orbit of the restricted actions a representative. The sets of representatives are denoted by $\mathcal{R}_4$ and $\mathcal{R}_5$. We have $|\mathcal{R}_4| = |\mathcal{R}_5| = 10$. All elements in $\mathcal{B}(G, [2, 4, 6])$ are contained in two orbits of the Hurwitz action. We denote by A_1 and A_2 two representatives of these orbits. There are two disjoint pairs in $\{A_1, A_2\} \times \mathcal{R}_4$, and two disjoint pairs in $\{A_1, A_2\} \times \mathcal{R}_5$. According to 4.4, we have at least one representative for each $\mathfrak{T}$-orbit. Using the function"Orbi" we can identify the pairs above, which are $\mathfrak{T}$-equivalent. We find two equivalence classes. To verify the above computations a source code can be found at www.staff.uni-bayreuth.de/~bt300503/ in the file *script.txt*.

Remark 5.8 From the output files we can see that there is exactly one occurrence with $T_1(S) = T_2(S)$ and $n \geq 2$. In this case $G = G(128, 36)$. The types are $T_i = [4, 4, 4]$ and $n = 2$. We denote by (A_1, A_2) and (B_1, B_2) the representatives for the $\mathfrak{T}$-orbits from the output file *loop2.txt*. It remains to check if (A_2, A_1) and (B_1, B_2) are in the same $\mathfrak{T}$-orbit. A MAGMA computation shows that this is not the case. The source code for this computation is available at www.staff.uni-bayreuth.de/~bt300503/ in the file *script.txt*.

Now we can give our main theorem, which implies in particular Theorem 1.1.

Theorem 5.9 *Let $S = (C_1 \times C_2)/G$ be a regular surface isogenous to a product of unmixed type with $\chi(\mathcal{O}_S) = 2$. Then $g(C_1)$, $g(C_2)$, the group G and the corresponding types $T_1(S)$, $T_2(S)$ are:*

$g(C_1)$	$g(C_2)$	G	Id	$T_1(S)$	$T_2(S)$	n
17	43	$PSL(2, \mathbb{F}_7) \times \mathbb{Z}_2$	$\langle 336, 209 \rangle$	$[2, 3, 14]$	$[4^3]$	2
49	9	$(\mathbb{Z}_2)^3 \rtimes_\varphi \mathfrak{S}_4$	$\langle 192, 955 \rangle$	$[2^2, 4^2]$	$[2, 4, 6]$	2
49	8	$PSL(2, \mathbb{F}_7)$	$\langle 168, 42 \rangle$	$[7^3]$	$[3^2, 4]$	2
17	22	$PSL(2, \mathbb{F}_7)$	$\langle 168, 42 \rangle$	$[3^2, 7]$	$[4^3]$	2
5	81	$(\mathbb{Z}_2)^4 \rtimes_\varphi \mathcal{D}_5$	$\langle 160, 234 \rangle$	$[2, 4, 5]$	$[4^4]$	5
17	17	$G(128, 36)$	$\langle 128, 36 \rangle$	$[4^3]$	$[4^3]$	2
9	31	$\mathfrak{S}_5$	$\langle 120, 34 \rangle$	$[2, 5, 6]$	$[2^2, 4^2]$	1
5	49	$(\mathbb{Z}_2)^4 \rtimes_\varphi \mathcal{D}_3$	$\langle 96, 195 \rangle$	$[2, 4, 6]$	$[4^4]$	1
25	9	$(\mathbb{Z}_2)^4 \rtimes_\varphi \mathcal{D}_3$	$\langle 96, 227 \rangle$	$[2^5]$	$[3, 4^2]$	1
9	17	$(\mathbb{Z}_2)^3 \rtimes_\varphi \mathcal{D}_4$	$\langle 64, 73 \rangle$	$[2^3, 4]$	$[2^2, 4^2]$	1
9	17	$U(4, 2)$	$\langle 64, 138 \rangle$	$[2^3, 4]$	$[2^2, 4^2]$	1
13	11	$\mathcal{A}_5$	$\langle 60, 5 \rangle$	$[5^3]$	$[2^2, 3^2]$	1
41	4	$\mathcal{A}_5$	$\langle 60, 5 \rangle$	$[3^5]$	$[2, 5^2]$	2
9	16	$\mathcal{A}_5$	$\langle 60, 5 \rangle$	$[3, 5^2]$	$[2^5]$	2
5	31	$\mathcal{A}_5$	$\langle 60, 5 \rangle$	$[3^2, 5]$	$[2^6]$	1
5	25	$\mathfrak{S}_4 \times \mathbb{Z}_2$	$\langle 48, 48 \rangle$	$[2^3, 3]$	$[4^4]$	1
9	13	$\mathfrak{S}_4 \times \mathbb{Z}_2$	$\langle 48, 48 \rangle$	$[2^3, 6]$	$[2^2, 4^2]$	1
13	9	$\mathfrak{S}_4 \times \mathbb{Z}_2$	$\langle 48, 48 \rangle$	$[2^5]$	$[4^2, 6]$	1
3	49	$\mathfrak{S}_4 \times \mathbb{Z}_2$	$\langle 48, 48 \rangle$	$[2, 4, 6]$	$[2^8]$	2
9	9	$(\mathbb{Z}_2)^3 \rtimes_\varphi \mathbb{Z}_4$	$\langle 32, 22 \rangle$	$[2^2, 4^2]$	$[2^2, 4^2]$	1
9	9	$\mathcal{D}_4 \times (\mathbb{Z}_2)^2$	$\langle 32, 46 \rangle$	$[2^5]$	$[2^5]$	1
17	5	$(\mathbb{Z}_2)^4 \rtimes_\varphi \mathbb{Z}_2$	$\langle 32, 27 \rangle$	$[2^3, 4^2]$	$[2^3, 4]$	3
9	9	$(\mathbb{Z}_2)^4 \rtimes_\varphi \mathbb{Z}_2$	$\langle 32, 27 \rangle$	$[2^2, 4^2]$	$[2^5]$	1
5	13	$\mathfrak{S}_4$	$\langle 24, 12 \rangle$	$[2^2, 3^2]$	$[4^4]$	1
3	25	$\mathfrak{S}_4$	$\langle 24, 12 \rangle$	$[3, 4^2]$	$[2^8]$	1
5	9	$\mathcal{D}_4 \times \mathbb{Z}_2$	$\langle 16, 11 \rangle$	$[2^5]$	$[2^3, 4^2]$	1
9	5	$(\mathbb{Z}_2)^2 \rtimes_\varphi \mathbb{Z}_4$	$\langle 16, 3 \rangle$	$[2^3, 4^2]$	$[2^2, 4^2]$	2
9	5	$(\mathbb{Z}_2)^4$	$\langle 16, 14 \rangle$	$[2^6]$	$[2^5]$	2
3	17	$\mathcal{D}_4 \times \mathbb{Z}_2$	$\langle 16, 11 \rangle$	$[2^3, 4]$	$[2^8]$	2
7	4	$(\mathbb{Z}_3)^2$	$\langle 9, 2 \rangle$	$[3^5]$	$[3^4]$	1
5	5	$(\mathbb{Z}_2)^3$	$\langle 8, 5 \rangle$	$[2^6]$	$[2^6]$	1
3	9	$(\mathbb{Z}_2)^3$	$\langle 8, 5 \rangle$	$[2^5]$	$[2^8]$	1

Each row in the table corresponds to a union of connected components of the Gieseker moduli space of surfaces of general type with $K_S^2 = 16$ and $= 2$. The number n of these components is given in the last column.

Acknowledgments The author thanks Ingrid Bauer and Sascha Weigl for several suggestions, useful discussions and very careful reading of the paper.

References

1. I. Bauer, F. Catanese, F. Grunewald, The classification of surfaces with $p_g = q = 0$ isogenus to a product. Pure Appl. Math. Q. part 1 **4**(2), part 1, 547–586 (2008)
2. I. Bauer, F. Catanese, F. Grunewald, R. Pignatelli, Quotients of product of curves, new surfaces with $p_g = 0$ and their fundamental groups. Am. J. Math. **134**(4), 993–1049 (2012)
3. A. Beauville, L' Inégalité $p_g \geq 2q - 4$ pour les surfaces de type général. Bull. Soc. Math. **110**, 343–346 (1982)
4. A. Beauville, *Complex Algebraic Surfaces*. vol. 68, London Mathematical Society Lecture Note Series (Cambridge University Press, Cambridge, 1983)
5. T. Breuer, *Characters and Automorphism Groups of Compact Riemann Surfaces*. London Mathematical Society Lecture Note Series, vol. 280 (Cambridge University Press, Cambridge, 2000)
6. W. Burnside, On groups of order $p^a q^b$. Proc. Lond. Math. Soc. 388 (1904)
7. G. Carnovale, F. Polizzi, The classification of surfaces with $p_g = q = 1$ isogenus to a product of curves. Adv. Geom. **9**(2), 233–256 (2009)
8. F. Catanese, Fibred surfaces, varieties isogenus to a product and related moduli spaces. Am. J. Math. **122**, 1–44 (2000)
9. F. Catanese, C. Ciliberto, M. Mendes Lopes, On the classification of irregular surfaces of general type with nonbirational bicanonical map. Trans. Am. Math. Soc. **350**, 275–308 (1998)
10. A. Garbagnati, M. Penegini, K3 surfaces with a non-symplectic automorphism and product-quotient surfaces with cyclic groups (2013). arXiv:1303.1653v1
11. D. Gieseker, Global moduli for surfaces of general type. Invent. Math. **43**(3), 233–282 (1977)
12. P.A. Griffiths, Variations on a theorem of Abel. Invent. Math. **35**, 321–390 (1976)
13. P. Griffiths, J. Harris, *Principles of Algebraic Geometry*. A Wiley-interscience publication (Wiley, New York, 1978)
14. C.D. Hacon, R. Pardini, Surfaces with $p_g = q = 3$. Trans. Amer. Math. Soc. **354**(7), 2631–2638 (2002)
15. MAGMA Databases of Groups; http://magma.maths.usyd.edu.au/magma/handbook/text/668
16. R. Miranda, *Algebraic Curves and Riemann Surfaces*. Graduate Studies in Mathematics, vol. 5 (American Mathematical Society, Providence, 1995)
17. M. Penegini, The classification of isotrivially fibred surfaces with $p_g = q = 2$, and topics on Beauville surfaces. PhD thesis, Universität Bayreuth, 2010
18. M. Penegini, The classification of isotrivially fibred surfaces with $p_g = q = 2$. Collect. Math. **62**(3), 239–274 (2011)
19. G.P. Pirola, Surfaces with $p_g = q = 3$. Manuscripta Math. **108**(2), 163–170 (2002)

Characteristically Simple Beauville Groups, II: Low Rank and Sporadic Groups

Gareth A. Jones

Abstract A Beauville surface is a rigid complex surface of general type, isogenous to a higher product by the free action of a finite group, called a Beauville group. Here we consider which characteristically simple groups can be Beauville groups. We show that if G is a cartesian power of a simple group $L_2(q)$, $L_3(q)$, $U_3(q)$, $Sz(2^e)$, $R(3^e)$, or of a sporadic simple group, then G is a Beauville group if and only if it has two generators and is not isomorphic to A_5.

MSC classification: 20B25 (primary) $\cdot$ 14J50 $\cdot$ 20G40 $\cdot$ 20H10 (secondary)

1 Introduction

A Beauville surface $\mathcal{S}$ of unmixed type is a complex algebraic surface which is isogenous to a higher product (that is, $\mathcal{S} = (\mathcal{C}_1 \times \mathcal{C}_2)/G$, where $\mathcal{C}_1$ and $\mathcal{C}_2$ are complex algebraic curves of genus $g_i \geq 2$, and G is a finite group acting freely on their product), and is rigid in the sense that G preserves the factors $\mathcal{C}_i$, with $\mathcal{C}_i/G \cong \mathbb{P}^1(\mathbb{C})$ $(i = 1, 2)$ and the induced covering $\beta_i : \mathcal{C}_i \to \mathbb{P}^1(\mathbb{C})$ ramified over three points. (In this paper we will not consider Beauville surfaces of mixed type, where elements of G transpose the factors $\mathcal{C}_i$.) The first examples, with $\mathcal{C}_1 = \mathcal{C}_2$ Fermat curves, were introduced by Beauville [3, p. 159] in 1978. Subsequently, these surfaces have been intensively studied by geometers such as Bauer et al. and Catanese (see [1, 2, 6], for instance).

Recently, there has been considerable interest in determining which groups G, known as Beauville groups, can arise in this way. It is easily shown that the alternating group A_5 is not a Beauville group. In 2005 Bauer et al. [1] conjectured that every other non-abelian finite simple group is a Beauville group. In 2010, more or less simultaneously, Garion et al. [13] proved that this is true with finitely many possible

G.A. Jones ($\boxtimes$)
School of Mathematics, University of Southampton, Southampton SO17 1BJ, UK
e-mail: G.A.Jones@maths.soton.ac.uk

© Springer International Publishing Switzerland 2015

I. Bauer et al. (eds.), *Beauville Surfaces and Groups*, Springer Proceedings in Mathematics & Statistics 123, DOI 10.1007/978-3-319-13862-6_7

exceptions, while Guralnick and Malle [14] and Fairbairn et al. [11] proved that it is completely true. It is natural to consider other classes of finite groups, and this paper is part of a project to extend this result to characteristically simple groups, those with no characteristic subgroups other than 1 and G.

A finite group is characteristically simple if and only if it is isomorphic to a cartesian power H^k of a simple group H. In [6], Catanese showed that the abelian Beauville groups are those isomorphic to C_n^2 for some n coprime to 6; these are characteristically simple if and only if n is prime, so we may assume that H is a non-abelian finite simple group.

The ramification condition on the coverings β_i implies that a Beauville group is a quotient of a triangle group (in two ways), so it is a 2-generator group. Clearly we require that $G \not\cong A_5$, and in [20] it was conjectured that these two conditions are also sufficient:

Conjecture *Let G be a non-abelian finite characteristically simple group. Then G is a Beauville group if and only if it is a 2-generator group not isomorphic to A_5.*

This was proved in [20] in the case where $G = H^k$ with H an alternating group A_n. (Further details on the background to this problem are given there, while [19] gives a more general survey of Beauville groups.) By the classification of finite simple groups, this leaves the simple groups of Lie type, together with the 26 sporadic simple groups. Our aim here is to verify the conjecture for five families of groups of small Lie rank, specifically the projective special linear groups $L_2(q)$ and $L_3(q)$, the unitary groups $U_3(q)$, the Suzuki groups $Sz(2^e)$ and the 'small' Ree groups $R(3^e)$, together with the sporadic simple groups (see [8, 33] for notation and properties of these and other finite simple groups).

The 2-generator condition on H^k can be restated as follows. For any 2-generator finite group H, there is an integer $c_2(H)$ such that H^k is a 2-generator group if and only if $k \le c_2(H)$. If H is a non-abelian finite simple group (and hence a 2-generator group, by the classification), then $c_2(H)$ is equal to the number $d_2(H)$ of orbits of Aut H on ordered pairs of generators of H. Thus in this case, H^k is a 2-generator group if and only if $k \le d_2(H)$.

For the groups $L_2(q) = SL_2(q)/\{\pm I\}$, which are simple for all prime powers $q \ge 4$, with $L_2(4) \cong L_2(5) \cong A_5$, we will prove:

Theorem 1.1 *Let $G = H^k$, where $k \ge 1$ and $H = L_2(q)$ for a prime power $q \ge 4$. Then the following are equivalent:*

(a) G is a Beauville group;
(b) G is a 2-generator group not isomorphic to A_5;
(c) $k \le d_2(H)$, and if $q \le 5$ then $k > 1$.

The groups $L_3(q)$ and $U_3(q)$ are simple for all prime powers $q \ge 2$ and 3 respectively, while the Suzuki groups $Sz(2^e)$ and the Ree groups $R(3^e)$ are simple for all odd $e \ge 3$. For these and for the sporadic simple groups, we have a slightly simpler formulation:

Theorem 1.2 *Let $G = H^k$, where $k \geq 1$ and H is a simple group $L_3(q)$, $U_3(q)$, $Sz(2^e)$ or $R(3^e)$, or a sporadic simple group. Then the following are equivalent:*

(a) G is a Beauville group;
(b) G is a 2-generator group;
(c) $k \leq d_2(H)$.

In Theorems 1.1 and 1.2, the implications (a) $\Rightarrow$ (b) $\Rightarrow$ (c) are straightforward (see Sects. 2 and 3), and the main task is to prove that (c) implies (a). The case where $k = 1$ having been dealt with by others (see above), it is thus sufficient to prove the following:

Theorem 1.3 *If H is a simple group $L_2(q)$, $L_3(q)$, $U_3(q)$, $Sz(2^e)$ or $R(3^e)$, or a sporadic simple group, then H^k is a Beauville group for each $k = 2, \ldots, d_2(H)$.*

It is hoped that the methods developed here and in [20] may allow this result to be extended to all finite simple groups of Lie type, thus proving the above conjecture.

The results proved here and in [20] yield further examples of Beauville groups, as follows. Let us say that two groups are *orthogonal* if they have no non-identity epimorphic images in common. It is straightforward to show that a cartesian product of finitely many mutually orthogonal Beauville groups is a Beauville group: orthogonality ensures that the obvious triples generate the group. It follows that if $G_1, \ldots, G_r$ are characteristically simple Beauville groups, which are cartesian powers of mutually non-isomorphic simple groups, then $G_1 \times \cdots \times G_r$ is also a Beauville group.

2 Background and Method of Proof

The proof of Theorem 1.3 is based on the characterisation by Bauer et al. [1, 2] of Beauville groups as those finite groups G such that

- G is a smooth quotient of two hyperbolic triangle groups

$$\Delta_i = \Delta_i(l_i, m_i, n_i) = \langle A_i, B_i, C_i \mid A_i^{l_i} = B_i^{m_i} = C_i^{n_i} = A_i B_i C_i = 1 \rangle$$

for $i = 1, 2$,
- $\Sigma_1 \cap \Sigma_2 = 1$, where Σ_i is the set of conjugates of powers of the elements a_i, b_i, c_i of G which are images of the canonical generators A_i, B_i, C_i of Δ_i.

Here 'smooth' means that a_i, b_i and c_i have orders l_i, m_i and n_i, so that the kernel K_i of the epimorphism $\Delta_i \to G$ is torsion-free, and thus a surface group. 'Hyperbolic' means that $l_i^{-1} + m_i^{-1} + n_i^{-1} < 1$, so that Δ_i acts on the hyperbolic plane $\mathbb{H}$. The isomorphism $G \cong \Delta_i/K_i$ induces an action of G as a group of automorphisms of the compact Riemann surface (or complex algebraic curve) $\mathcal{C}_i = \mathbb{H}/K_i$ of genus $g_i > 1$, with the natural projection $\mathcal{C}_i \to \mathcal{C}_i/G \cong \mathbb{H}/\Delta_i \cong \mathbb{P}^1(\mathbb{C})$ branched over three points (corresponding to the fixed points of A_i, B_i and C_i in $\mathbb{H}$). Since Σ_i is the set of elements of G with fixed points in C_i, the condition $\Sigma_1 \cap \Sigma_2 = 1$ is necessary and sufficient for G to act freely on $C_1 \times C_2$, so that $S := (C_1 \times C_2)/G$ is a complex surface. The generating triples (a_i, b_i, c_i) for G (always satisfying $a_i b_i c_i = 1$) are said to have *type* (l_i, m_i, n_i), and the pair of them form a *Beauville structure* of type $(l_1, m_1, n_1; l_2, m_2, n_2)$ in G.

Since Σ_i is closed under taking powers, the condition $\Sigma_1 \cap \Sigma_2 = 1$ is equivalent to the condition that

$$\Sigma_1^{(p)} \cap \Sigma_2^{(p)} = \emptyset$$

for every prime p dividing $l_1 m_1 n_1$ and $l_2 m_2 n_2$, where $\Sigma_i^{(p)}$ denotes the set of elements of order p in Σ_i. In many cases, this condition is simpler to verify.

If $G = H^k$ for some H then the members of any generating triple for G are k-tuples $a_i = (a_{ij})$, $b_i = (b_{ij})$, $c_i = (c_{ij})$ such that (a_{ij}, b_{ij}, c_{ij}) is a generating triple for H for each $j = 1, \ldots, k$, and these k triples are mutually inequivalent under the action of Aut H. If H is a non-abelian finite simple group then conversely any k mutually inequivalent generating triples for H yield a generating triple for G. Moreover, if $G \not\cong A_5$ this triple is hyperbolic. To prove that G is a Beauville group it is therefore sufficient to find two k-tuples of mutually inequivalent generating triples for H, and then to show that the resulting pair of generating triples for G satisfy $\Sigma_1^{(p)} \cap \Sigma_2^{(p)} = \emptyset$ for all primes p dividing $|H|$.

There are k-tuples of mutually inequivalent generating triples for H if and only if $k \leq d_2(H)$. In [20], suitable generating triples for the simple groups $H = A_n$ were exhibited by explicitly defining their members a_{ij}, b_{ij} and c_{ij}, and then applying results on groups containing such permutations to show that they generate H. Here we adopt a similar approach for the groups $H = L_2(q)$, using Dickson's description of their maximal subgroups [9, Ch. XII] to show that various specific triples generate H. However, for the remaining simple groups it is more convenient to use a character-theoretic formula of Frobenius (see Sect. 6) to prove the existence of triples of various types, and then again to use knowledge of the maximal subgroups of H to show that they generate H.

In each case, the condition that $\Sigma_1^{(p)} \cap \Sigma_2^{(p)} = \emptyset$ for all p is ensured by carefully choosing the two k-tuples of generating triples for H so that powers of a_1, b_1 or c_1 of order p have different supports in $\{1, 2, \ldots, k\}$ from those of a_2, b_2 or c_2. Several results which guarantee this were proved in [20], and for completeness their proofs are outlined in Sect. 4. These results are then applied to the groups $L_2(q)$ in Sect. 5, and to the other families of simple groups in Sects. 7–10. However, before doing this we will briefly discuss in Sect. 3 the values of k for which H^k is a 2-generator group.

3 Generating Cartesian Powers

If H is a finite group, let

$$c_2(H) = \max\{k \in \mathbb{N} \mid H^k \text{ is a 2-generator group}\}.$$

If H^k is a Beauville group then it is a 2-generator group, and thus a quotient of the free group F_2 of rank 2, so

$$k \le c_2(H) \le d_2(H)$$

where $d_2(H)$ denotes the number of normal subgroups N of F_2 with $F_2/N \cong H$. Following Hall [15], let $\phi_2(H)$ be the number of 2-bases (ordered generating pairs) for H. Any 2-base for H determines an epimorphism $\theta : F_2 \to H$, and hence a normal subgroup $N = \ker \theta$ of F_2 with $F_2/N \cong H$; conversely, every such normal subgroup arises in this way, with two 2-bases corresponding to the same normal subgroup N if and only if they are equivalent under an automorphism of H. Thus $d_2(H)$ is equal to the number of orbits of $\mathrm{Aut}\, H$ on 2-bases for H. Since only the identity automorphism can fix a 2-base, this action is semiregular, so we have the following:

Lemma 3.1 *If H is a finite group then*

$$d_2(H) = \frac{\phi_2(H)}{|\mathrm{Aut}\, H|}.$$

If H is a non-abelian finite simple group, and F_2 has k normal subgroups with quotient H, then their intersection is a normal subgroup with quotient H^k. It follows then that $c_2(H) = d_2(H)$ for such a group H, so we have the following:

Corollary 3.2 *If H is a non-abelian finite simple group and H^k is a Beauville group, then*

$$k \le c_2(H) = d_2(H) = \frac{\phi_2(H)}{|\mathrm{Aut}\, H|}.$$

Since generating triples (a, b, c) for a group (with $abc = 1$) correspond bijectively to its 2-bases (a, b), we have the following useful characterisation of generating triples for H^k:

Corollary 3.3 *Let H be a non-abelian finite simple group. Then k-tuples $a = (a_j)$, $b = (b_j)$ and $c = (c_j)$ form a generating triple for H^k if and only if their components (a_j, b_j, c_j) for $j = 1, \ldots, k$ form k mutually inequivalent generating triples for H.* $\qquad \square$

By results of Dixon [10], Kantor and Lubotzky [22], and Liebeck and Shalev [25], a randomly-chosen pair of elements of a non-abelian finite simple group H generate the whole group with probability approaching 1 as $|H| \to \infty$, so

$$d_2(H) \sim \frac{|H|^2}{|\text{Inn } H|.|\text{Out } H|} = \frac{|H|}{|\text{Out } H|} \quad \text{as} \quad |H| \to \infty.$$

For each infinite family of such groups, $|\text{Out } H|$ is constant or grows much more slowly than $|H|$, so $d_2(H)$ grows almost as fast as $|H|$. In particular, as $q = p^e \to \infty$ with p prime, the infinite families considered here satisfy

- $d_2(L_2(q)) \sim q^3/d^2 e$ while $|L_2(q)| \sim q^3/d$, where $d = \gcd(2, q - 1)$,
- $d_2(L_3(q)) \sim q^8/2d^2 e$ while $|L_3(q)| \sim q^8/d$, where $d = \gcd(3, q - 1)$,
- $d_2(U_3(q)) \sim q^7/2d^2 e$ while $|U_3(q)| \sim q^7/d$, where $d = \gcd(3, q + 1)$,
- $d_2(Sz(q)) \sim q^5/e$ while $|Sz(q)| \sim q^5$,
- $d_2(R(q)) \sim q^7/e$ while $|R(q)| \sim q^7$.

One can illustrate how close the asymptotic estimate $|H|/|\text{Out } H|$ is to $d_2(H)$ as follows. For any finite group H we have

$$\phi_2(H) = \left|H^2 \setminus \bigcup_M M^2\right| \geq |H|^2 - \sum_M |M|^2,$$

where M ranges over the maximal subgroups of H. If H is perfect then each such M has $|H : M|$ conjugates, so

$$\phi_2(H) \geq |H|^2\left(1 - \sum_{i=1}^{r} \frac{1}{|H : M_i|}\right),$$

where M_i ranges over a set of representatives of the r conjugacy classes of maximal subgroups of H. It follows that if H also has trivial centre (and in particular if H is a non-abelian finite simple group) then

$$1 \geq \frac{d_2(H)}{|H|/|\text{Out } H|} \geq 1 - \sum_{i=1}^{r} \frac{1}{|H : M_i|}.$$

When H is large, the sum of the right is typically very small. Thus, for the simple groups $H = L_2(q)$, $U_3(q)$, $Sz(q)$ and $R(q)$ we have

$$\sum_{i=1}^{r} \frac{1}{|H : M_i|} \sim \frac{1}{q^m} \quad \text{as} \quad q \to \infty,$$

where $m = 1, 3, 2$ and 3 respectively, with the sum dominated by the term corresponding to the doubly transitive permutation representation of H of degree $q^m + 1$; for $H = L_n(q)$ with $n \geq 3$ the sum grows like $2/q^{n-1}$, corresponding to the two doubly transitive representations of degree $(q^n - 1)/(q - 1)$, on the points and hyperplanes of $\mathbb{P}^{n-1}(q)$.

The situation is similar for the sporadic simple groups: for instance, if H is O'Nan's group $O'N$, with $r = 13$ conjugacy classes of maximal subgroups, we have

$$\frac{|H|}{|\text{Out } H|} = \frac{460815505920}{2} = 230407752960$$

and

$$\sum_{i=1}^{r} \frac{1}{|H : M_i|} = 0.00001726863378\ldots,$$

so

$$230407752960 \geq d_2(H) \geq 230403774132.$$

4 Beauville Structures in Cartesian Powers

Corollary 3.3 describes the generating triples in a cartesian power $G = H^k$ of a non-abelian finite simple group H, in terms of those for H. To avoid confusion, we will use notations such as $\mathcal{T} = (a, b, c)$ and $T = (x, y, z)$ for generating triples in G and H respectively. We now consider sufficient conditions for two triples $\mathcal{T}_i = (a_i, b_i, c_i)$ in G to satisfy $\Sigma_1^{(p)} \cap \Sigma_2^{(p)} = \emptyset$ for all primes p, so that they form a Beauville structure for G.

If $g = (g_j) \in \Sigma_i^{(p)}$ for some prime p, then g has order p and is conjugate in G to a power of some $d_i = a_i, b_i$ or c_i, so for each $j \in \mathbb{N}_k := \{1, \ldots, k\}$ its jth coordinate g_j has order 1 or p. The support

$$\text{supp}(g) = \{j \in \mathbb{N}_k \mid g_j \neq 1\}$$

of g is then the p-*summit* $S_p(d_i)$ of d_i, defined to be the set of $j \in \mathbb{N}_k$ for which the power of p dividing the order of its jth coordinate d_{ij} is greatest. The p-summit $S_p(\mathcal{T})$ of a triple $\mathcal{T} = (a, b, c)$ in G is the set $\{S_p(a), S_p(b), S_p(c)\}$ of subsets of $\mathbb{N}_k$. The following is obvious:

Lemma 4.1 *Let H be any finite group. If two triples $\mathcal{T}_i = (a_i, b_i, c_i)$ $(i = 1, 2)$ in $G = H^k$ have disjoint p-summits $S_p(\mathcal{T}_i)$ for some prime p, then $\Sigma_1^{(p)} \cap \Sigma_2^{(p)} = \emptyset$.* $\square$

An element of H is p-*full* if its order is divisible by the highest power of p dividing the exponent $\exp(H)$ of H. If d_i has at least one p-full coordinate d_{ij}, then

$$S_p(d_i) = F_p(d_i) := \{j \in \mathbb{N}_k \mid d_{ij} \text{ is } p\text{-full}\}.$$

If each element d_i of a triple $\mathcal{T}_i$ in G has a p-full coordinate, it is easier to determine $S_p(\mathcal{T}_i)$, and hence to ensure that pairs of such triples $\mathcal{T}_i$ satisfy the assumptions of Lemma 4.1.

Given a triple $T = (x, y, z)$ in H, let $v_p(T)$ be the number of p-full elements among x, y and z. Two triples T_i ($i = 1, 2$) in H are *p-distinguishing* if $v_p(T_1) \neq v_p(T_2)$, and *strongly p-distinguishing* if, in addition, whenever $v_p(T_i) = 0$ then either p^2 does not divide $\exp(H)$ or p does not divide any of the three periods of T_i.

Lemma 4.2 *Suppose that a non-abelian finite simple group H has a set $\{(T_{1,s}, T_{2,s}) \mid s = 1, \ldots, t\}$ of ordered pairs $(T_{1,s}, T_{2,s})$ of generating triples such that*

1. *for each prime p dividing $|H|$ there is some $s = s(p) \in \{1, \ldots, t\}$ such that $T_{1,s}$ and $T_{2,s}$ are a strongly p-distinguishing pair;*
2. *for each $i = 1, 2$ the $3t$ triples consisting of $T_{i,1}, \ldots, T_{i,t}$ and their cyclic permutations are mutually inequivalent.*

Then $G := H^k$ is a Beauville group for each $k = 3t, \ldots, d_2(H)$.

Proof Let $T_{i,s} = (x_{i,s}, y_{i,s}, z_{i,s})$ for each $i = 1, 2$ and $s = 1, \ldots, t$. Since $3t \leq k \leq d_2(H)$, for each i one can form a generating triple $\mathcal{T}_i = (a_i, b_i, c_i)$ for G by using k inequivalent generating triples for H in the different coordinate positions, with $T_{i,s}$ and its two cyclic permutations in the jth positions where $j = 3s - 2, 3s - 1$ or $3s$ for $s = 1, \ldots, t$.

Suppose that $g \in \Sigma_1^{(p)} \cap \Sigma_2^{(p)}$ for some prime p. There is a strongly p-distinguishing pair $(T_{1,s}, T_{2,s})$, with $v_p(T_{1,s}) \neq v_p(T_{2,s})$. First suppose that $v_p(T_{i,s}) > 0$ for $i = 1, 2$, so that each element $d_i = a_i, b_i$ or c_i of $\mathcal{T}_i$ has at least one p-full coordinate, and hence $S_p(d_i) = F_p(d_i)$. It follows that

$$|\mathrm{supp}(g) \cap \{3s - 2, 3s - 1, 3s\}| = v_p(T_{i,s}) \tag{1}$$

for $i = 1$ and 2, which is impossible since $v_p(T_{1,s}) \neq v_p(T_{2,s})$. Otherwise, we may suppose without loss of generality that $v_p(T_{1,s}) > 0 = v_p(T_{2,s})$, with (1) satisfied for $i = 1$ but not for $i = 2$. Some $d_2 = a_2, b_2$ or c_2 must then have a non-p-full coordinate of order divisible by p in position $3s - 2, 3s - 1$ or $3s$, which is impossible if the periods of $T_{2,s}$ are coprime to p or if p^2 does not divide $\exp(H)$, as we are assuming. Thus $\Sigma_1^{(p)} \cap \Sigma_2^{(p)} = \emptyset$, so the triples $\mathcal{T}_i$ form a Beauville structure for G. $\qquad\square$

If we simply assume that the pairs $T_{1,s}$ and $T_{2,s}$ are p-distinguishing, we have the following rather weaker conclusion:

Lemma 4.3 *Suppose that a non-abelian finite simple group H has a set $T = \{(T_{1,s}, T_{2,s}) \mid s = 1, \ldots, t\}$ of ordered pairs $(T_{1,s}, T_{2,s})$ of generating triples such that*

1. *for each prime p dividing $|H|$ there is some $s = s(p) \in \{1, \ldots, t\}$ such that $T_{1,s}$ and $T_{2,s}$ are a p-distinguishing pair;*
2. *the $6t$ triples consisting of the $2t$ triples $T_{i,s}$ and their cyclic permutations are mutually inequivalent.*

Then $G := H^k$ is a Beauville group for each $k = 6t, \ldots, d_2(H)$.

Proof The proof is similar. The first $3t$ coordinates of a_i, b_i and c_i are defined as before, but now those in positions $j = 3t+1, \ldots, 6t$ are the coordinates of a_{3-i}, b_{3-i} or c_{3-i} in positions $j - 3t$. Each of the six generators a_i, b_i, c_i of G now has at least one p-full coordinate for each prime p dividing $|H|$, so any $g \in \Sigma_1^{(p)} \cap \Sigma_2^{(p)}$ satisfies (1) for $i = 1$ and 2, leading to a contradiction as before. $\square$

For the small values of k omitted by these lemmas one can often use the following:

Lemma 4.4 *Let H be a non-abelian finite simple group with a set of $r \geq 2$ mutually inequivalent generating triples of type (l, m, n), where l, m and n are mutually coprime. Then $G := H^k$ is a Beauville group for each $k = 2, \ldots, 6r$.*

Proof If the specified generating triples for H are (x_j, y_j, z_j) for $j = 1, \ldots, r$, then one can form $6r$ mutually inequivalent generating triples for H by cyclically permuting each (x_j, y_j, z_j) and each $(z_j^{-1}, y_j^{-1}, x_j^{-1})$. If $k = 2, \ldots, 6r$ one can form two generating triples

$$a_1 = (x_1, x_2, \ldots), \quad b_1 = (y_1, y_2, \ldots), \quad c_1 = (z_1, z_2, \ldots)$$

and

$$a_2 = (x_1, y_2, \ldots), \quad b_2 = (y_1, z_2, \ldots), \quad c_2 = (z_1, x_2, \ldots),$$

for G, using k of these triples in the different coordinate positions. Since l, m and n are mutually coprime, any prime p dividing l, m or n divides exactly one of them, so if $g \in \Sigma_1^{(p)}$ then $|\mathrm{supp}(g) \cap \{1, 2\}| = 2$ whereas if $g \in \Sigma_2^{(p)}$ then $|\mathrm{supp}(g) \cap \{1, 2\}| = 1$. Thus $\Sigma_1^{(p)} \cap \Sigma_2^{(p)} = \emptyset$, so these two triples form a Beauville structure for G. $\square$

5 The Groups $L_2(q)$

We will now apply the results in the preceding section to show that the simple groups $L_2(q)$ satisfy Theorem 1.3.

Let $G = H^k$ where $H = L_2(q)$ with $q = p_0^e \geq 4$ for some prime p_0, and with $k = 2, \ldots, d_2(H)$. Since $L_2(4) \cong L_2(5) \cong A_5$ and $L_2(9) \cong A_6$, the main theorem of [20], which proves Theorem 1.3 for $H = A_n$ ($n \geq 5$), allows us to assume that $q = 7$ or 8 or $q \geq 11$. We will use Lemma 4.2 with $t = 1$ to show that G is a Beauville group. However, this lemma does not apply when $k = 2$, so we will deal with this case first by a separate argument.

Let $k = 2$. Results of Macbeath [26] show that there exist generating triples (x_i, y_i, z_i) ($i = 1, 2$) for H of types (q_1, q_1, p_0) and (q_2, q_2, p_0), where $q_1, q_2 = (q \pm 1)/d$ and $d = \gcd(2, q - 1)$. For instance, in the equation

$$\begin{pmatrix} s & 1 \\ -1 & 0 \end{pmatrix} \begin{pmatrix} 0 & 1 \\ -1 & t \end{pmatrix} = \begin{pmatrix} -1 & s+t \\ 0 & -1 \end{pmatrix}$$

in $SL_2(q)$ one can choose s and t to be the traces of elements x_i and y_i of order $o(x_i) = o(y_i) = q_i$ in H for $i = 1, 2$, with the matrix on the right representing an element $z_i^{-1} = x_i y_i$ of order p_0 provided $s + t \neq 0$. Dickson's description of the subgroups of $L_2(q)$ (see [9, Ch. XII]) shows that the only maximal subgroups containing elements of orders q_i and p_0 are dihedral groups of order q_i when $p_0 = 2$, impossible here since x_i and y_i do not commute, and stabilisers of points in $\mathbb{P}^1(q)$, also impossible since z_i fixes only ∞, which is not fixed by x_i. Thus each triple (x_i, y_i, z_i) generates H. Having different types, these two generating triples are mutually inequivalent, so we obtain a generating triple

$$a_1 = (x_1, x_2), \quad b_1 = (y_1, y_2), \quad c_1 = (z_1, z_2)$$

for G. Similarly G has a second generating triple

$$a_2 = (x_2, z_1), \quad b_2 = (y_2, x_1), \quad c_2 = (z_2, y_1).$$

The primes p dividing $|H| = q(q^2 - 1)/d$ are p_0 and those dividing q_1 or q_2. Since p_0, q_1 and q_2 are mutually coprime, if $p = p_0$ then any $g \in \Sigma_i^{(p)}$ has $|\mathrm{supp}(g)| = 2$ or 1 for $i = 1$ or 2 respectively, so $\Sigma_1^{(p)} \cap \Sigma_2^{(p)} = \emptyset$. If $g \in \Sigma_i^{(p)}$ with p dividing q_1, then $\mathrm{supp}(g) = \{1\}$ or $\{2\}$ for $i = 1$ or 2, while for p dividing q_2 it is the other way round. Thus $\Sigma_1^{(p)} \cap \Sigma_2^{(p)} = \emptyset$ for all p, so H^2 is a Beauville group.

Now let $k = 3, \ldots, d_2(H)$. Let (x_i, y_i, z_i) $(i = 1, 3)$ be generating triples for H of types (q_1, q_1, p_0) and (q_1, q_2, q_2) for H. For instance, the first could be as above in the case $k = 2$; the second could consist of the images in H of matrices in $SL_2(q)$ of the form

$$\begin{pmatrix} u & 1 \\ -1 & 0 \end{pmatrix}, \quad \begin{pmatrix} v & w \\ 0 & v^{-1} \end{pmatrix} \quad \text{and} \quad \begin{pmatrix} -w & -uw - v^{-1} \\ v & uv \end{pmatrix},$$

where u is the trace of an element of H of order q_1, v generates the multiplicative group of the field $\mathbb{F}_q$, and w is chosen so that $uv - w$ is the trace of an element of H of order q_2. Again, it follows from [9, Ch. XII] that this triple generates H.

We now apply Lemma 4.2, with $t = 1$, to this pair of triples. Again, we must consider the primes p dividing $|H|$. The generator z_1 is p_0-full, since the Sylow p_0-subgroups of H are elementary abelian, while x_1, y_1 and x_3 are p-full for all primes p dividing q_1, and y_3 and z_3 are p-full for primes dividing q_2. The hypotheses of Lemma 4.2 are satisfied, with $t = 1$, so H^k is a Beauville group for each $k = 3, \ldots, d_2(H)$.

6 Counting Triples

For the other families of finite simple groups we shall consider, it is easier to prove the existence of generating triples of various types by means of the following enumerative formula of Frobenius [12] than to exhibit them directly as we did for $H = L_2(q)$.

Proposition 6.1 *If X, Y and Z are conjugacy classes in a finite group H, then the number $v_H(X, Y, Z)$ of triples $(x, y, z) \in X \times Y \times Z$ such that $xyz = 1$ is given by*

$$v_H(X, Y, Z) = \frac{|X| \cdot |Y| \cdot |Z|}{|H|} \sum_{\chi} \frac{\chi(x)\chi(y)\chi(z)}{\chi(1)},$$

where $x \in X$, $y \in Y$, $z \in Z$, and χ ranges over the irreducible complex characters of H.

Proofs of this and related results can be found in [18, Appendix] or [30, Theorem 7.2.1]. By summing over all choices of conjugacy classes X, Y and Z of elements of given orders l, m and n, one can find the number $v_H(l, m, n)$ of triples of type (l, m, n) in H.

An equivalent, and sometimes more convenient, version of Proposition 6.1 is that

$$v_H(X, Y, Z) = \frac{|H|^2}{|C_H(x)| \cdot |C_H(y)| \cdot |C_H(z)|} S(X, Y, Z),$$

where $S(X, Y, Z)$ denotes the character sum

$$S(X, Y, Z) = \sum_{\chi} \frac{\chi(x)\chi(y)\chi(z)}{\chi(1)}.$$

In most of the cases we shall consider, one finds that many characters χ take the value 0 on at least one of the three classes X, Y and Z, so they do not contribute to this sum. Even when a non-principal character χ contributes, the absolute values $|\chi|$ on these classes are often so much smaller than the degree $\chi(1)$ that the sum is dominated by the contribution, equal to 1, from the principal character, so that $S(X, Y, Z) \approx 1$. These estimates often show that $S(X, Y, Z) > 0$, so that $v_H(X, Y, Z) > 0$ and hence H has such triples (x, y, z).

We are particularly interested in generating triples, those not contained in any maximal subgroup M of H. The number $\phi_H(l, m, n)$ of these of type (l, m, n) clearly satisfies

$$\phi_H(l, m, n) \geq v_H(l, m, n) - \sum_{M} v_M(l, m, n),$$

where M ranges over the maximal subgroups of M. In particular, if $v_M(l, m, n) = 0$ for all such M then

$$\phi_H(l, m, n) = v_H(l, m, n).$$

7 The Suzuki Groups

We will apply this method first to the Suzuki groups $Sz(q)$, using the description of their conjugacy classes, subgroups and characters given by Suzuki in [31] (see also [33, Sect. 4.2]).

If $q = 2^e$ for some odd $e \geq 3$, then the group $H = Sz(q) = {}^2B_2(q)$ is a simple group of order $q^2(q^2 + 1)(q - 1)$. It has four conjugacy classes of maximal cyclic subgroups, of mutually coprime orders 4, $q_1 = q + \sqrt{2q} + 1$, $q_2 = q - \sqrt{2q} + 1$ and $q_3 = q - 1$. (Note that $q_1 q_2 = q^2 + 1$.) It follows that any prime p dividing $|H|$ divides precisely one of these four integers, and any non-identity element of G is conjugate to an element of precisely one of these four subgroups. In particular, a pair of generating triples of types $(4, q_2, q_3)$ and (q_1, q_3, q_3), if they exist, will satisfy the conditions of Lemma 4.2 with $t = 1$, and hence prove that H^k is a Beauville group for $k = 3, \ldots, d_2(H)$.

To prove that such triples exist, we use Frobenius's formula (Proposition 6.1), together with the character table of H in [31, Sect. 17]. Conjugacy classes X, Y and Z of elements of orders 4, q_2 and q_3 satisfy $|X| = 4(q^2+1)(q-1)$, $|Y| = q^2 q_1(q-1)$ and $|Z| = q^2(q^2+1)$, so each choice of such classes gives rise to $4q_1|H| > 0$ triples $(x, y, z) \in X \times Y \times Z$ with $xyz = 1$. (Only the principal character χ makes a non-zero contribution to the sum in Proposition 6.1, all other characters vanishing on at least one of the chosen classes.) The subgroups of H are described in [31, Sect. 15]; no proper subgroup contains elements of orders q_2 and q_3, so each of these triples generates H. A similar argument applies to triples of type (q_1, q_3, q_3): each appropriate choice of conjugacy classes yields $q_2(q + 1)|H|$ generating triples, with only the irreducible characters of degrees 1 and q^2 contributing to the sum.

The case $k = 2$ is covered by Lemma 4.4: since $|\mathrm{Aut}\, H| = e|H|$ and $4q_1 > 4q > e$ there are at least two inequivalent generating triples of type $(4, q_2, q_3)$, and their periods are mutually coprime. Thus the Suzuki groups satisfy Theorem 1.3.

8 The Small Ree Groups

Similar methods can be applied to the small Ree groups $H = R(q) = {}^2G_2(q)$, where $q = 3^e$ for some odd $e \geq 3$. (These groups, described by Ree in [28], are called 'small' to distinguish them from the 'large' Ree groups ${}^2F_4(2^e)$.) Each of these groups H is simple, of order $q^3(q^3 + 1)(q - 1)$. The Sylow 2-subgroups of H are elementary abelian, of order 8; there is a single conjugacy class of involutions, with centralisers isomorphic to $C_2 \times L_2(q)$, of order $q(q^2 - 1)$. The Sylow 3-subgroups are non-abelian, of order q^3 and exponent 9; there are three conjugacy classes of elements of order 9, all with centralisers of order $3q$. There are cyclic Hall subgroups A_i of mutually coprime odd orders $q_i := (q - 1)/2$, $(q + 1)/4$, $q - \sqrt{3q} + 1$ and $q + \sqrt{3q} + 1$ for $i = 0, \ldots, 3$. (Note that $q \equiv 3 \bmod (8)$ and $q^3 + 1 = (q + 1)(q^2 - q + 1) = 4q_1 q_2 q_3$.)

We will again use Lemmas 4.2 and 4.4, now applied to triples of types $(2, q_1, q_3)$ and $(9, q_0, q_2)$, with six mutually coprime periods. For each prime p dividing $|H|$, one of these triples contains a single p-full element, while the elements of the other triple all have orders coprime to p. Provided they exist, such triples therefore satisfy the hypotheses of these two lemmas.

One can count triples by applying Frobenius's formula to the character table for H given by Ward in [32]. In the case of triples of type $(9, q_0, q_2)$, there are respectively three, $\varphi(q_0)/2$ and $\varphi(q_2)/6$ conjugacy classes of elements of orders 9, q_0 and q_2, denoted by YT^i ($i = 0, \pm 1$), R^a and V in [32], with centralisers of orders $3q$, $q - 1$ and q_2. Only the principal character makes a non-zero contribution to the character sum in the formula, so the number of triples of this type in H is

$$3 \cdot \frac{\varphi(q_0)}{2} \cdot \frac{\varphi(q_2)}{6} \cdot \frac{|H|^2}{3q(q-1)q_2} = \frac{1}{12}\varphi(q_0)\varphi(q_2)q^5(q^3 + 1)(q - 1)^2 q_3 > 0.$$

A similar calculation applies to triples of type $(2, q_1, q_3)$. The elements of orders $2, q_1$ and q_3, denoted by J, S^a and W in [32], form 1, $\varphi(q_1)/6$ and $\varphi(q_3)/6$ conjugacy classes, with centralisers of orders $q(q^2 - 1)$, $q + 1$ and q_3 respectively. The character sum takes the value

$$1 + \frac{q \cdot (-1)^2}{q^3} + 2 \cdot \frac{q_0 \cdot (-1) \cdot 1}{q_0 q_2 \sqrt{q/3}} = 1 + \frac{1}{q^2} - \frac{2}{q_2\sqrt{q/3}},$$

with ξ_3 of degree q^3, and ξ_6 and ξ_8 of degree $q_0 q_2 \sqrt{q/3}$ the only non-principal characters contributing to it. It follows that the number of triples of this type is

$$\frac{\varphi(q_1)}{6} \cdot \frac{\varphi(q_3)}{6} \cdot \frac{|H|^2}{q(q^2 - 1)(q + 1)q_3}\left(1 + \frac{1}{q^2} - \frac{2}{q_2\sqrt{q/3}}\right) > 0.$$

The maximal subgroups of H have been determined by Levchuk and Nuzhin [24] and Kleidman [23] (see also [33, Sect. 4.5.3]). They are as follows:

- the point stabilisers in the doubly transitive permutation representation of degree $q^3 + 1$, or equivalently, the normalisers of the Sylow 3-subgroups, of order $q^3(q - 1)$;
- the centralisers of involutions, of order $q(q^2 - 1)$;
- the normalisers of Hall subgroups A_i ($i = 1, 2, 3$), of order $6q_i$;
- subgroups isomorphic to $R(3^f)$ where e/f is prime.

None of these subgroups contains elements of orders q_0 and q_2, or of orders q_1 and q_3, so each triple of type $(9, q_0, q_2)$ or $(2, q_1, q_3)$ generates H. Dividing the above numbers of triples by $|\mathrm{Aut}\, H| = e|H|$ gives the numbers of equivalence classes of triples of these two types. As in the case of the Suzuki groups, Lemmas 4.2 and 4.4 now show that these Ree groups satisfy Theorem 1.3.

9 The Sporadic Simple Groups

In this section we will prove that the 26 sporadic simple groups satisfy Theorem 1.3. Most of the information we need about these groups can be found in [8] or [33]; in particular, we have adopted the notation of [8] for conjugacy classes, maximal subgroups, etc.

9.1 The Mathieu Groups

It is convenient to deal with the Mathieu groups together, in view of the inclusion relations between them. Each of the Mathieu groups $H = M_n$ ($n = 11, 12, 22, 23$ or 24) has order divisible by at most six primes. This allows us to use Lemma 4.2 with $t = 1$ to deal with the values $k = 3, \ldots, d_2(H)$, while Lemma 4.4 deals with $k = 2$. In each case we can use Frobenius's formula (Proposition 6.1), applied to the character tables in [8], to count triples of a given type, and the lists of maximal subgroups in [8] to eliminate those which do not generate H. Here, as with some of the other sporadic simple groups, we will compute the exact numbers of equivalence classes of generating triples of the required types, even though it is usually sufficient to show that at least one or two exist: this is because it may be useful in other contexts (such as regular maps and hypermaps) to know the number of normal subgroups of a given triangle group with quotient isomorphic to H. For instance, by the calculation in Sect. 9.2.1 there are, up to isomorphism, 18 orientably regular maps of type $\{5, 19\}$ with orientation-preserving automorphism group isomorphic to Janko's group J_1; the Riemann-Hurwitz formula shows that they all have genus 21,715.

M_{11}

Let $H = M_{11}$, a simple group of order $2^4.3^3.5.11$ and exponent $2^3.3.5.11$ with $|\mathrm{Out}\, H| = 1$. Frobenius's formula and the character table in [8] show that there are $2^6.3^3.5^2.11 = 20|H| = 20|\mathrm{Aut}\, H|$ triples of type $(3, 8, 11)$ in H; they all generate H since no maximal subgroup contains elements of orders 8 and 11, so they form 20 equivalence classes.

Similarly, there are $54|\mathrm{Aut}\, H|$ triples of type $(5, 11, 11)$. The only maximal subgroups containing elements of order 5 and 11 are the twelve isomorphic to $L_2(11)$. Applying Frobenius's formula to $L_2(11)$ shows that $8|\mathrm{Aut}\, H|$ of these triples are contained in such maximal subgroups, so the remainder form 46 equivalence classes of generating triples.

A pair of generating triples of types $(3, 8, 11)$ and $(5, 11, 11)$ satisfy the hypotheses of Lemma 4.2, so H^k is a Beauville group for each $k = 3, \ldots, d_2(H)$. Since there are at least two inequivalent generating triples of type $(3, 8, 11)$, Lemma 4.4 proves this for $k = 2$.

M_{12}

The group $H = M_{12}$, of order $2^6.3^3.5.11$ with $|\mathrm{Out}\, H| = 2$, has the same exponent as M_{11}, so we can use the same method, with triples of the same types $(3, 8, 11)$ and $(5, 11, 11)$, provided they generate H. Frobenius's formula shows that H has $96|\mathrm{Aut}\, H|$ triples of type $(3, 8, 11)$. The only maximal subgroups containing elements of orders 8 and 11 are the 24 isomorphic to M_{11}; we have already counted such triples in M_{11}, so we find that there are $20|\mathrm{Aut}\, H|$ non-generating triples, and hence 76 equivalence classes of generating triples of this type.

Similarly there are $158|\mathrm{Aut}\, H|$ triples of type $(5, 11, 11)$ in H. The only maximal subgroups containing elements of orders 5 and 11 are the 24 isomorphic to M_{11} and the 144 isomorphic to $L_2(11)$. Having already counted triples of this type in these two groups, we find that there are $54|\mathrm{Aut}\, H|$ and $4|\mathrm{Aut}\, H|$ triples contained in maximal subgroups respectively isomorphic to M_{11} and to $L_2(11)$. These two sets of triples are disjoint, since any triple of type $(5, 11, 11)$ contained in $L_2(11)$ generates that group, so there are $58|\mathrm{Aut}\, H|$ non-generating triples, and hence 100 equivalence classes of generating triples of this type. From this point, as with the remaining Mathieu groups, the argument follows that for M_{11}.

M_{22}

The group $H = M_{22}$ has order $2^7.3^2.5.7.11$ and exponent $2^3.3.5.7.11$ with $|\mathrm{Out}\, H| = 2$. There are $2^{17}.3^3.5.7.11 = 1536|\mathrm{Aut}\, H|$ triples of type $(5, 11, 11)$ in H. The only maximal subgroups containing elements of order 11 are the $2^5.3.7 = 672$ isomorphic to $L_2(11)$. We have seen that $L_2(11)$ contains $2^5.3.5.11 = 5280$ such triples, and is generated by each of them, so there are $2^{10}.3^2.5.7.11 = 4|\mathrm{Aut}\, H|$ non-generating triples and hence 1532 equivalence classes of generating triples of this type.

Similarly, there are $216|\mathrm{Aut}\, H|$ triples of type $(3, 7, 8)$ in H, each generating H or contained in a maximal subgroup isomorphic to $AGL_3(2)$. There are 330 such maximal subgroups, each containing 224 elements of order 3 (eight in each of the 28 two-point stabilisers, isomorphic to S_4, in the natural affine action of this group), and 384 elements of order 7 (48 in each of the eight point stabilisers, isomorphic to $L_3(2)$). Thus there are at most $330 \times 224 \times 384 = 32|\mathrm{Aut}\, H|$ non-generating triples and hence at least 184 equivalence classes of generating triples of type $(3, 7, 8)$ in H.

M_{23}

The group $H = M_{23}$ has order $2^7.3^2.5.7.11.23$ and exponent $2^3.3.5.7.11.23$ with $|\mathrm{Out}\, H| = 1$. There are $2^{10}.3^2.5.7^2.11^2.23 = 616|\mathrm{Aut}\, H|$ triples of type $(3, 8, 23)$. The only maximal subgroups with elements of order 23 are the normalisers of Sylow 23-subgroups, of order 11.23, so each of these triples generates H, giving 616 equivalence classes.

There are $2^{15}.3^3.5.7.11.23^2 = 17{,}664|\text{Aut } H|$ triples of type $(5, 7, 11)$ in H. The only maximal subgroups with elements of orders $5, 7$ and 11 are the 23 subgroups isomorphic to M_{22}. Now there are $2^{16}.3^4.5.7.11$ triples of type $(5, 7, 11)$ in M_{22}, each generating M_{22}, so H has $2^{16}.3^4.5.7.11.23 = 4608|\text{Aut } H|$ non-generating triples and hence 13,056 equivalence classes of generating triples of this type.

M_{24}

The group $H = M_{24}$ has order $2^{10}.3^3.5.7.11.23$ and $|\text{Out } H| = 1$, with the same exponent $2^3.3.5.7.11.23$ as M_{23}, so we can again use triples of types $(3, 8, 23)$ and $(5, 7, 11)$. There are $2^{11}.3^2.5.7.11.23.769 = 1538|\text{Aut } H|$ triples of type $(3, 8, 23)$ in H with the generator of order 3 in the class $3A$, and $2^{14}.3^3.5.7.11^2.17.23 = 1992|\text{Aut } H|$ with it in class $3B$, giving a total of $3530|\text{Aut } H|$ triples. The maximal subgroups of H have been determined by Choi [7]. The only maximal subgroups with elements of orders 8 and 23 are the 24 isomorphic to M_{23}. As we have seen, M_{23} has $2^{10}.3^2.5.7^2.11^2.23$ triples of type $(3, 8, 23)$, each generating M_{23}, so there are $2^{13}.3^3.5.7^2.11^2.23 = 616|\text{Aut } H|$ non-generating triples of this type in H, and hence 2914 equivalence classes of generating triples.

There are $22{,}256|\text{Aut } H|$ triples of type $(5, 7, 11)$ in H. The only maximal subgroups with elements of orders $5, 7$ and 11 are the 24 isomorphic to M_{23} and the $\binom{24}{2} = 276$ isomorphic to $M_{22} : 2 = \text{Aut } M_{22}$. These are the stabilisers of points and of unordered pairs of points in the natural representation of H. Any triple of type $(5, 7, 11)$ in $\text{Aut } M_{22}$ must be contained in the subgroup M_{22} of index 2 fixing two points, so it is contained in two of the point stabilisers. However, no such triple lies in three point stabilisers, since elements of order 11 fix just two points. We have seen that there are $2^{15}.3^3.5.7.11.23^2 = 17{,}664|\text{Aut } M_{23}| = 17{,}664|\text{Aut } H|/24$ triples of type $(5, 7, 11)$ in M_{23}, and $2^{16}.3^4.5.7.11 = 192|\text{Aut } H|/23$ in M_{22}, so the number of generating triples of this type is

$$22{,}256|\text{Aut } H| - 17{,}664|\text{Aut } H| + 276 \cdot \frac{192|\text{Aut } H|}{23} = 6896|\text{Aut } H|,$$

and they form 6896 equivalence classes.

9.2 Other Small Sporadic Simple Groups

The method applied to the Mathieu groups can also be applied to some of the other sporadic simple groups H, provided they are 'small' in the sense that not too many primes divide $|H|$. In these cases, it is sufficient to show that there are inequivalent generating triples satisfying the hypotheses of Lemmas 4.2 and 4.4. We will deal with these groups in ascending order of magnitude.

The Janko Group J_1

The first Janko group $H = J_1$ has order $2^3.3.5.7.11.19$ and exponent $2.3.5.7.11.19$ with $|\text{Out}\,H| = 1$. We will use triples of types $(2, 5, 19)$ and $(3, 7, 11)$. There are $2^4.3^2.5.7.11.19 = 18|\text{Aut}\,H|$ triples of type $(2, 5, 19)$ in H. They are all generating triples, since no maximal subgroup (classified by Janko [17]) contains elements of orders 5 and 19, so they form 18 equivalence classes. Similarly there are $2^5.3.5.7.11.19^2 = 76|\text{Aut}\,H|$ triples of type $(3, 7, 11)$ in H. No maximal subgroup contains elements of orders 7 and 11, so these form 76 equivalence classes of generating triples.

The Janko Group J_2

The second Janko group $H = J_2$ has order $2^7.3^3.5^2.7$ and exponent $2^3.3.5.7$ with $|\text{Out}\,H| = 2$. We will use triples of types $(3, 5, 7)$ and $(7, 7, 8)$. There are $2^{16}.3^4.5^2.7 = 768|\text{Aut}\,H|$ triples of type $(7, 7, 8)$ in H, with only the principal character and that of degree 225 contributing to the character sum. The only triples of this type contained in proper subgroups are those which generate one of the 100 maximal subgroups isomorphic to $U_3(3)$. This group contains $2^{12}.3^3.7$ such triples, so there are $2^{14}.3^3.5^2.7 = 64|\text{Aut}\,H|$ non-generating triples in H, and hence there are 704 equivalence classes of generating triples of this type.

Any triple of type $(3, 5, 7)$ generates H since no maximal subgroup has elements of orders 5 and 7. For simplicity we will consider only those triples for which the element of order 3 is in the Aut H-invariant conjugacy class $3B$, consisting of those with a centraliser of order 36, and the element of order 5 is in class $5C$ or $5D$, transposed by Aut H, with a centraliser of order 50. We find that there are 49 equivalence classes of such triples, with only the principal character and that of degree 288 appearing in the character sum.

The Higman-Sims Group HS

The Higman-Sims group $H = HS$ has order $2^9.3^2.5^3.7.11$ and exponent $2^3.3.5.7.11$ with $|\text{Out}\,H| = 2$. We will use triples of types $(7, 11, 11)$ and $(8, 11, 15)$. The only maximal subgroups containing elements of order 11 are isomorphic to M_{11} or M_{22}; neither of these groups has elements of order 15, and only M_{22} has elements of order 7. There are $1,04,760|\text{Aut}\,H|$ triples of type $(7, 11, 11)$ in H, with only the principal character and that of degree 3200 contributing to the character sum. There are 100 subgroups isomorphic to M_{22}, in total containing (and generated by) $2048|\text{Aut}\,H|$ such triples, so there are $1,02,712$ equivalence classes of triples of this type. There are $50,496|\text{Aut}\,H|$ triples of type $(8, 11, 15)$, with only the principal character and that of degree 175 contributing to the character sum. Each of these triples generates H, so they form $50,496$ equivalence classes.

The McLaughlin Group $M^c L$

The McLaughlin group $H = M^c L$ has order $2^7.3^6.5^3.7.11$ and exponent $2^3.3^2.5.7.11$ with $|\mathrm{Out}\, H| = 2$. We will use triples of types $(5, 7, 11)$ and $(8, 9, 9)$. In counting triples of type $(5, 7, 11)$ we will consider only those for which the element of order 5 is in the conjugacy class $5A$, with centralisers of order 750. The only maximal subgroups with elements of orders 7 and 11 are those isomorphic to M_{22}; the permutation character of H on the cosets of such a subgroup, as given in [8], takes the value 0 on elements of $5A$, so these elements do not lie in such subgroups. Thus each of these triples generates H. There are $2^{12}.3^{11}.5^3.7.11 = 3888|\mathrm{Aut}\, H|$ of them, with only the principal character contributing to the character sum, so they form 3888 equivalence classes.

There are $2^{13}.3^6.5^6.7^2.11^2 = 2^5.5^3.7.11|\mathrm{Aut}\, H| = 3,08,000|\mathrm{Aut}\, H|$ triples of type $(8, 9, 9)$ in H, again with only the principal character contributing to the character sum. The only maximal subgroups containing elements of orders 8 and 9 are those isomorphic to $U_4(3)$, $3^{1+4}_+:2S_5$ or $3^4:M_{10}$; in the last two cases, any triple of this type must lie in the subgroup $3^{1+4}_+:2A_5$ or $3^4:A_6$ of index 2, whereas this has no elements of order 8. Thus any non-generating triple of this type must lie in a maximal subgroup isomorphic to $U_4(3)$. There are 275 of these, any two of them intersecting in a subgroup isomorphic to $3^4:A_8$ or $L_3(4)$; these have no elements of orders 8 or 9 respectively, so each such triple lies in a unique maximal subgroup. Since $U_4(3)$ has $2^{14}.3^7.5.7.23$ triples of type $(8, 9, 9)$ (with only the principal character and the two of degree 35 contributing to the character sum), there are $2^{14}.3^7.5^3.7.11.23 = 2^7.5.7|\mathrm{Aut}\, H| = 4480|\mathrm{Aut}\, H|$ non-generating triples in H, and hence $3,03,584|\mathrm{Aut}\, H|$ generating triples, forming $3,03,584$ equivalence classes.

One can apply this method to a few other sporadic simple groups, such as J_3, He, Ru, $O'N$ and Co_3, but it is more efficient to deal with them by using a separate method.

9.3 The Larger Sporadic Simple Groups

In the case of the larger sporadic simple groups H, and in particular those of order divisible by more than six primes, Lemma 4.2 may not be applicable, since it may be impossible to find a pair of generating triples satisfying its hypotheses for each of the primes dividing $|H|$. Instead, one can use Lemma 4.3 with $t > 1$ to deal with values $k \geq 6t$, and Lemma 4.4 to deal with small values of k. In each case, we find two mutually coprime integers q_1 and q_2 dividing $|H|$ (usually the largest two prime divisors) such that no maximal subgroup of H contains elements of orders q_1 and q_2. (The maximal subgroups have been completely classified in all cases except that of the Monster simple group M, and in this case sufficient is known about the maximal subgroups to justify this claim.) It follows that for any l, a triple of type (l, q_1, q_2)

in H must generate the whole group. Provided they exist, we can therefore use pairs of triples of types (l, q_1, q_2) for various l to distinguish the primes dividing $|H|$. By Lemma 4.3 this will deal with large values of k, and one of these types, with l coprime to q_1 and q_2, will also deal with small k by Lemma 4.4, provided there are sufficiently many inequivalent triples of that type.

As before, we can use Frobenius's formula (Proposition 6.1) to show that suitable triples exist. In all the relevant cases, inspection of the character table of H shows that the character sum $S(X, Y, Z)$ is very close to 1, with the modulus $|\chi(g)|$ of each non-principal character χ of H for g in each of the chosen conjugacy classes X, Y and Z significantly smaller than the degree $\chi(1)$. Thus $\nu_H(X, Y, Z) > 0$, so suitable generating triples exist, and moreover in sufficient numbers for Lemma 4.4 to deal with the small values of k not covered by Lemma 4.3.

We will give a detailed explanation of this method as it applies to the largest and the smallest of the remaining sporadic simple groups, namely the Monster group and the third Janko group. For the other groups, in order to omit tedious and repetitive numerical details we will simply refer to Table 1, which lists the groups in ascending order of magnitude, together with their exponents and the choices of q_1 and q_2.

Table 1 The larger sporadic simple groups

H	Name	Exponent	q_1, q_2
J_3	Janko	$2^3.3^2.5.17.19$	17, 19
He	Held	$2^5.3.5.7.17$	7, 17
Ru	Rudvalis	$2^3.3.5.7.13.29$	13, 29
Suz	Suzuki	$2^3.3^2.5.7.11.13$	11, 13
$O'N$	O'Nan	$2^4.3.5.7.11.19.31$	19, 31
Co_3	Conway	$2^4.3^2.5.7.11.23$	9, 23
Co_2	Conway	$2^4.3^2.5.7.11.23$	9, 23
Fi_{22}	Fischer	$2^4.3^2.5.7.11.13$	11, 13
HN	Harada-Norton	$2^3.3^2.5^2.7.11.19$	11, 19
Ly	Lyons	$2^3.3^2.5^2.7.11.31.37.67$	37, 67
Th	Thompson	$2^3.3^3.5.7.13.19.31$	19, 31
Fi_{23}	Fischer	$2^4.3^3.5.7.11.13.17.23$	17, 23
Co_1	Conway	$2^4.3^2.5.7.11.13.23$	13, 23
J_4	Janko	$2^4.3.5.7.11.23.29.37.43$	37, 43
Fi'_{24}	Fischer	$2^4.3^3.5.7.11.13.17.23.29$	23, 29
B	Baby monster	$2^5.3^3.5^2.7.11.13.17.19.23.31.47$	31, 47
M	Monster	$2^5.3^3.5^2.7.11.13.17.19.23.29.31.41.47.59.71$	59, 71

The Monster Group M

The Monster group $H = M$ has order $2^{46}.3^{20}.5^9.7^6.11^2.13^3.17.19.23.29.31.41.47.$ 59.71 and exponent $2^5.3^3.5^2.7.11.13.17.19.23.29.31.41.47.59.71$, with $|\text{Out } H| = 1$. We will apply the method described above, with $q_1 = 59$ and $q_2 = 71$. The maximal subgroups of M are not yet completely known, but it is known that the only maximal subgroups of order divisible by 59 or 71 are those isomorphic to $L_2(59)$ or $L_2(71)$ (see [33, Table 5.6]); thus no proper subgroup has order divisible by both 59 and 71, so a triple of type $(l, 59, 71)$ for any l must generate H. Provided they exist, we can therefore use a pair of triples of types $(59, 59, 71)$ and $(59, 71, 71)$ to distinguish the primes 59 and 71, and six pairs of triples of types $(l, 59, 71)$, where $l = 11, 13, 19, 23, 25, 27, 29, 31, 32, 41, 47, 119$ $(=7.17)$, for the remaining thirteen primes $p < 59$ dividing $|H|$. By Lemma 4.3 this will deal with values $k \geq 42$, and one of the latter types, with mutually coprime periods, will cover smaller k by Lemma 4.4. Inspection of the character table of H in [8] shows that for each of the above types one can choose appropriate conjugacy classes X, Y and Z so that the character sum $S(X, Y, Z)$ is very close to 1. This shows that such generating triples exist, and in sufficient numbers for Lemma 4.4 to deal with the cases $k = 2, \ldots, 41$.

In fact, all but 14 of the 193 non-principal irreducible characters χ vanish on elements of order 59 or 71, so they do not contribute to any of the character sums, while the non-vanishing characters, of degrees $\chi(1) \geq 8980616927734375$, all take values $\chi(g) = \pm 1$ or $(-1 \pm i\sqrt{59})/2$ on those classes. The character values on the other classes needed are also much smaller than these degrees: for instance, if g has order $l = 11$ then $|\chi(g)| \leq 190$ for all χ. Since there are two conjugacy classes each of elements of orders 59 and 71, and one of order 11, with centralisers of orders $59, 71$ and 1045440 respectively, it follows that the number of equivalence classes of generating triples of type $(11, 59, 71)$ is approximately

$$4 \times \frac{|H|}{59.71.1045440} \approx 7 \cdot 38 \times 10^{44}.$$

The estimates for the other types required are even larger, since there are more elements of order l in the cases where $l > 11$.

The Janko Group J_3

The third Janko group $H = J_3$ has order $2^6.3^5.5.17.19$ and exponent $2^3.3^2.5.17.19$ with $|\text{Out } H| = 2$. Although there are only five primes dividing $|H|$, it is more convenient to apply the method used for M than that used for the Mathieu groups. We will take $q_1 = 17$ and $q_2 = 19$. The only maximal subgroups of H with elements of order 19 are those isomorphic to $L_2(19)$; this group has no elements of order 17, so for any l, a triple of type $(l, 17, 19)$ must generate H. Provided they exist, one can therefore use a pair of triples of types $(17, 17, 19)$ and $(17, 19, 19)$ to distinguish the primes 17 and 19, a pair of types $(8, 17, 19)$ and $(9, 17, 19)$ for the primes $p = 2$

and 3, and a pair of types $(5, 17, 19)$ and (for instance) $(4, 17, 19)$ for $p = 5$. As in the case of M, the character values guarantee the existence of all the required triples: the only non-principal irreducible characters χ not vanishing on elements of orders 17 or 19, and hence contributing to any of the relevant character sums $S(X, Y, Z)$, are one of degree 324, two of degree 1215 and three of degree 1920; these satisfy $|\chi(g)| \leq 4$ for all elements g of order at least 4, so each $S(X, Y, Z)$ is very close to 1. By Lemma 4.3 this deals with the values $k = 18, \ldots, d_2(H)$. Frobenius's formula shows that there are, for instance, 21,312 equivalence classes of triples of type $(5, 17, 19)$, more than enough for Lemma 4.4 to deal with the values $k = 2, \ldots, 17$.

Table 1 shows how the same method can be applied to the remaining 15 sporadic simple groups. Thus each of the 26 sporadic simple groups satisfies Theorem 1.3, and the proof is complete.

10 The Groups $L_3(q)$ and $U_3(q)$

The method used for the larger sporadic simple groups can also be applied to other families of simple groups, such as the groups $L_3(q)$ and $U_3(q)$.

10.1 $L_3(q)$

The 3-dimensional projective special linear group $H = L_3(q)$ is simple for every prime power q. Since $L_3(2) \cong L_2(7)$ we may assume that $q > 2$. The group H contains elements h_1 and h_2 of orders

$$q_1 = \frac{q^2 + q + 1}{d} \quad \text{and} \quad q_2 = \frac{q^2 - 1}{d}$$

where $d = \gcd(q^2 + q + 1, 3) = \gcd(q - 1, 3)$. The maximal subgroups of H were determined for even q by Hartley [16], and for odd q first by Mitchell [27] and then (using more modern methods) by Bloom [5]: there are none containing elements of orders q_1 and q_2 (see also Kantor's classification [21] of linear groups containing Singer cycles). It follows that for any l, a triple of type (l, q_1, q_2) must generate H. The generic character table for H given by Simpson and Frame in [29] shows that $\chi(h_1)\chi(h_2) = 0$ for every non-principal irreducible character χ of H except the Steinberg character, of degree q^3; for this character we have $\chi(h_1) = 1 = -\chi(h_2)$, and $|\chi(h)| \leq q$ for all $h \neq 1$. Thus $S(X, Y, Z) > 0$ where X is any non-identity conjugacy class, and Y and Z are classes of elements of orders q_1 and q_2, so generating triples of type (l, q_1, q_2) exist whenever H contains elements of order $l \neq 1$.

Now

$$|H| = \frac{1}{d}q^3(q^2 + q + 1)(q + 1)(q - 1)^2,$$

so each prime divisor p of $|H|$ divides at least one of $q, q^2 + q + 1, q + 1$ and $q - 1$. These four integers are mutually coprime, except that $\gcd(q^2 + q + 1, q - 1) = 3$ if $q \equiv 1 \bmod (3)$, and $\gcd(q + 1, q - 1) = 2$ if q is odd. It follows from this that q_1 and q_2 are always mutually coprime. A pair of generating triples of types (q_1, q_1, q_2) and (q_1, q_2, q_2) can therefore be used to strongly distinguish all the primes dividing q_1 or q_2. This deals with every prime p dividing $|H|$, except the prime $p = p_0$ dividing q, and the prime $p = 3$ if $q \equiv 4$ or $7 \bmod (9)$ (so that q_1 and q_2 are both coprime to 3, whereas $|H|$ is not). The exponent q_0 of a Sylow p_0-subgroup of H is p_0 if $p_0 > 2$, and 4 if $p_0 = 2$, so a pair of triples of types (q_0, q_1, q_2) and $(3, q_1, q_2)$ will strongly distinguish p_0 if $p_0 \neq 3$, and will strongly distinguish 3 if $q \equiv 4$ or $7 \bmod (9)$; if $p_0 = 3$, so that $q \not\equiv 4$ or $7 \bmod (9)$, we can instead replace the second triple with one of type $(2, q_1, q_2)$ to strongly distinguish p_0. In either case these two pairs of triples satisfy the hypotheses of Lemma 4.2 with $t = 2$, so they deal with the values $k = 6, \ldots, d_2(H)$.

There are at least two inequivalent generating triples of type (p_0, q_1, q_2): for instance, one can choose the first element to be central or non-central in the unique Sylow p_0-subgroup containing it. These triples have mutually coprime periods, so Lemma 4.4(1) deals with $k = 2, \ldots, 12$, and hence the groups $H = L_3(q)$ satisfy Theorem 1.2.

10.2 $U_3(q)$

The proof for the unitary groups $H = U_3(q)$ is essentially the same. These groups have order

$$|H| = \frac{1}{d}q^3 \left(q^3 + 1\right)(q - 1) = \frac{1}{d}q^3 \left(q^2 - q + 1\right)(q + 1)(q - 1)$$

where $d = \gcd(q + 1, 3)$, and they are simple for all prime powers $q > 2$. They contain elements h_1 and h_2 of mutually coprime orders

$$q_1 = \frac{q^2 - q + 1}{d} \quad \text{and} \quad q_2 = \frac{q^2 - 1}{d},$$

all satisfying $H = \langle h_1, h_2 \rangle$ (see [33, Sect. 3.10.9] for the maximal subgroups of H, or [4] for subgroups containing a Singer cycle h_1). As with $L_3(q)$, the generic character table for H in [29] shows that generating triples of type (l, q_1, q_2) exist whenever H contains elements of order $l \neq 1$. A pair of generating triples of types (q_1, q_1, q_2) and (q_1, q_2, q_2) strongly distinguish each prime dividing $|H|$, apart from the prime p_0 dividing q, and the prime 3 if $q \equiv 2$ or $5 \bmod (9)$. The exponent

q_0 of a Sylow p_0-subgroup is as before, so for these primes one can use a pair of triples of types (q_0, q_1, q_2) and either $(3, q_1, q_2)$ or $(2, q_1, q_2)$ as $q \equiv 2, 5 \bmod (9)$ or not. Lemma 4.2, with $t = 2$, then deals with the values $k = 6, \ldots, d_2(H)$, and again Lemma 4.4(1), applied to triples of type (p_0, q_1, q_2), deals with small k. This completes the proof of Theorem 1.2.

References

1. I. Bauer, F. Catanese, F. Grunewald, Beauville surfaces without real structures I, in *Geometric Methods in Algebra and Number Theory*. Progress in Mathematics, vol. 235 (Birkhäuser, Boston, 2005), pp. 1–42
2. I. Bauer, F. Catanese, F. Grunewald, Chebycheff and Belyi polynomials, dessins d'enfants, Beauville surfaces and group theory. Mediterr. J. Math. **3**, 121–146 (2006)
3. A. Beauville, Surfaces algébriques complexes, Astérisque 54, *Soc. Math.* France, (1978)
4. A. Bereczky, Maximal overgroups of Singer elements in classical groups. J. Algebra **234**, 187–206 (2000)
5. D.M. Bloom, The subgroups of $PSL(3, q)$ for odd q. Trans. Am. Math. Soc. **127**, 150–178 (1967)
6. F. Catanese, Fibred surfaces, varieties isogenous to a product and related moduli spaces. Am. J. Math. **122**, 1–44 (2000)
7. C. Choi, On subgroups of M_{24}. II. The maximal subgroups of M_{24}. Trans. Am. Math. Soc. **167**, 29–47 (1972)
8. J.H. Conway, R.T. Curtis, S.P. Norton, R.A. Parker, R.A. Wilson, *ATLAS of Finite Groups* (Clarendon Press, Oxford, 1985)
9. L.E. Dickson, *Linear Groups* (Dover, New York, 1958)
10. J.D. Dixon, The probability of generating the symmetric group. Math. Z. **110**, 199–205 (1969)
11. B.T. Fairbairn, K. Magaard, C.W. Parker, Generation of finite simple groups with an application to groups acting on Beauville surfaces. Proc. London Math. Soc. **107**(3), 744–798 (2013) doi:10. 1112/plms/pds097
12. F.G. Frobenius, *Über Gruppencharaktere*, Sitzber. Königlich Preuss. Akad. Wiss. Berlin, pp. 985–1021 (1896)
13. S. Garion, M. Larsen, A. Lubotzky, Beauville surfaces and finite simple groups. J. Reine Angew. Math. **666**, 225–243 (2012)
14. R. Guralnick, G. Malle, Simple groups admit Beauville structures. J. London Math. Soc. **85**(2), 694–721 (2012)
15. P. Hall, The Eulerian functions of a group. Q. J. Math. **7**, 134–151 (1936)
16. R.W. Hartley, Determination of the ternary collineation groups whose coefficients lie in the $GF(2^n)$. Ann. Math. **27**(2), 140–158 (1925)
17. Z. Janko, A new finite simple group with abelian Sylow subgroups and its characterization. J. Algebra **3**, 147–186 (1966)
18. G.A. Jones, Enumeration of homomorphisms and surface-coverings. Q. J. Math. **46**, 485–507 (1995)
19. G.A. Jones, Beauville surfaces and groups: a survey, in *Rigidity and Symmetry*, ed. by R. Connelly, A. Weiss and W. Whiteley, *Fields Inst. Commun.***70**, pp. 205–225 (2014)
20. G.A. Jones, Characteristically simple Beauville groups, I: cartesian powers of alternating groups. in *Geometry, Groups and Dynamics*, ed. by C.S. Aravinda, W.M. Goldman, et al., Contemp. Math. **639**, to appear (2015)
21. W.M. Kantor, Linear groups containing a Singer cycle. J. Algebra **62**, 232–234 (1980)
22. W.M. Kantor, A. Lubotzky, The probability of generating a finite classical group. Geom. Dedicata **36**, 67–87 (1990)

23. P.B. Kleidman, The maximal subgroups of the Chevalley groups $G_2(q)$ with q odd, of the Ree groups ${}^2G_2(q)$, and of their automorphism groups. J. Algebra **117**, 30–71 (1988)
24. V.M. Levchuk, Ya. N. Nuzhin, Structure of Ree groups (Russian). Algebra Log. **24**, 26–41 (1985)
25. M.W. Liebeck, A. Shalev, The probability of generating a finite simple group. Geom. Dedicata **56**, 103–113 (1995)
26. A.M. Macbeath, Generators of the linear fractional groups, in *Number Theory (Houston 1967)*, ed. by W.J. Leveque, E.G. Straus (Am. Math. Soc., Providence, 1969), pp. 14–32. (Proc. Sympos. Pure Math. 12)
27. H.H. Mitchell, Determination of the ordinary and modular ternary linear groups. Trans. Am. Math. Soc. **12**, 207–242 (1911)
28. R. Ree, A family of simple groups associated with the simple Lie algebra of type (G_2). Am. J. Math. **83**, 432–462 (1961)
29. W.A. Simpson, J.S. Frame, The character tables for $SL(3, q)$, $SU(3, q^2)$, $PSL(3, q)$, $PSU(3, q^2)$. Can. J. Math. **25**, 486–494 (1973)
30. J.-P. Serre, *Topics in Galois Theory*, 2nd edn., A.K. Peters (Wellesley, 2008)
31. M. Suzuki, On a class of doubly transitive groups. Ann. Math. **75**(2), 105–145 (1962)
32. H.N. Ward, On Ree's series of simple groups. Trans. Am. Math. Soc. **121**, 62–89 (1966)
33. R.A. Wilson, *The Finite Simple Groups*. Graduate Texts in Mathematics, vol. 251 (Springer, London, 2009)

Remarks on Lifting Beauville Structures of Quasisimple Groups

Kay Magaard and Christopher Parker

Abstract In previous work, we developed theorems which produce a multitude of hyperbolic triples for finite classical groups. We apply these theorems to prove a conjecture of Bauer, Catanese and Grunewald, which asserts that all non-abelian finite quasisimple groups except for the alternating group of degree five are Beauville groups. Here we show that our results can be used to show that certain split- and Frattini extensions of quasisimple groups are also Beauville groups. We also discuss some open problems for future investigations.

2000 Mathematics Subject Classification. 20E34 · 20F05 · 14J29 · 30F10

1 Introduction

Suppose that G is a group. A *hyperbolic triple* in G is a triple $(x, y, z) \in G \times G \times G$ such that

$G = \langle x, y, z \rangle$ and $xyz = 1$; and
$1/o(x) + 1/o(y) + 1/o(z) < 1$.

A *Beauville structure* $\mathcal{B}$ of G is a pair of hyperbolic triples $((x_1, y_1, z_1), (x_2, y_2, z_2))$ in G such that no non-identity power of x_1, y_1 or z_1 is conjugate in G to a power of x_2, y_2 or z_2. A hyperbolic triple (x, y, z) has *type* $(o(x), o(y), o(z))$ and a Beauville structure $((x_1, y_1, z_1), (x_2, y_2, z_2))$ has *type*

$$((o(x_1), o(y_1), o(z_1)), (o(x_2), o(y_2), o(z_2))).$$

K. Magaard (✉) · C. Parker
School of Mathematics, University of Birmingham, Birmingham, B15 2TT
Edgbaston, UK
e-mail: k.magaard@bham.ac.uk

C. Parker
e-mail: c.w.parker@bham.ac.uk

© Springer International Publishing Switzerland 2015

I. Bauer et al. (eds.), *Beauville Surfaces and Groups*, Springer Proceedings in Mathematics & Statistics 123, DOI 10.1007/978-3-319-13862-6_8

In joint work with Fairbairn [2] the authors determined those quasisimple groups which have Beauville structures. The simple groups with Beauville structures have also been determined by Guralnick and Malle [4]. The main result of [2] asserts that, with the exception of $\mathrm{PSL}_2(5)$ and $\mathrm{SL}_2(5)$, all quasisimple groups do have such structures. Following on from these discoveries, a natural question that arises is

which extensions of the finite simple groups admit Beauville structures?

The purpose of this contribution is to point out some ways to begin to answer this question. One immediate restriction is that if G admits a Beauville structure, then G must be 2 generated. The recent conjecture of Gareth Jones [6, 7] suggests that in many interesting cases this might be sufficient.

One type of extension that generalizes perfect central extensions is called a *Frattini cover*. Particular amongst the Frattini covers are the non-split extensions of irreducible $\mathrm{GF}(p)G$-modules. In this note we draw attention to a corollary of a theorem of Guralnick and Tiep [5] which asserts that under certain mild restrictions Beauville structures for a group G lift to Frattini extensions of G. The details of this are in Sect. 2.

In Sect. 3, we consider split extensions $H = GV$ where V is a non-central elementary abelian minimal normal subgroup of H and G is a complement to V in H. We present an elementary criterion which guarantees that if G has a Beauville structure, then this structure lifts to H. Notice that a Beauville structure of GV may not project to a Beauville structure of G as the conjugacy condition may fail.

We choose examples of extensions of $\mathrm{SL}_d(p^a)$ to illustrate the type of results we find interesting and present only the most general statements. For instance, we present Beauville structures for p-Frattini covers $\mathrm{PSL}_d(p^a)$ for $d \geq 9$. The results we give are not the best results that can easily be obtained, but they are elementary and suffice to make our point. We also explain that if $G \cong \mathrm{SL}_d(p^a)$ and V is the natural $\mathrm{GF}(p^a)G$-module, then GV has a Beauville structure (again assuming that d is sufficiently large). Our final section poses a short series of questions which we hope will stimulate further research.

2 Frattini Covers of Beauville Groups

Suppose that G and X are groups such that $X/\Phi(X) \cong G$. Then X is called a *Frattini cover* of G. If $\Phi(X)$ is a p-group for some prime p, then X is a *p-Frattini cover.* By the Schur-Zassenhaus Theorem, if X is a p-Frattini cover of G and $\Phi(X) \neq 1$, then p divides $|G|$. The recent investigations in [5, 9] on p-Frattini covers are motivated by their connection with modular towers which were introduced by Michael Fried in connection with the inverse problem of Galois Theory.

Theorem 2.1 (Guralnick and Tiep [5]) *Let p be an odd prime and X be a p-Frattini cover of G. Assume that p does not divide the order of the Schur multiplier of G. Let $g_1, \ldots, g_r \in G$ satisfy*

  (i) $G = \langle g_1, \ldots, g_r \rangle$,

  (ii) $g_1 \ldots g_r = 1$, *and*

 (iii) *the order of each g_i is coprime to p.*

Then, for any $f \in \Phi(X)$, there exist $x_i \in X$, with $x_i \Phi(X) = g_i$ and $o(x_i) = o(g_i)$ such that $x_1 \ldots x_r = f$.

As an immediate corollary to the Guralnick-Tiep Theorem we obtain

Corollary 2.2 *Suppose that G has a Beauville structure $\mathcal{B}$ of type $((r, s, t), (u, v, w))$ and X is a p-Frattini cover of G. If p is coprime to $2rstuvw$ and to the order of the Schur multiplier of G, then X has a Beauville structure of type $((r, s, t), (u, v, w))$.*

Proof Take $f = 1$ in the Guralnick-Tiep Theorem. Then the hyperbolic triples in $\mathcal{B}$ lift to hyperbolic triples of X and provide a Beauville structure of X.   □

By Zsigmondy's Theorem, for a natural number $n > 2$ and a prime p such that $(n, p) \neq (6, 2)$, there is a prime which divides $p^n - 1$ which does not divide $p^k - 1$ for $k < n$. Such prime numbers are called *Zsigmondy primes*. Let $\zeta_{n,p}$ be a Zsigmondy prime dividing $p^n - 1$ but not $p^k - 1$ for $k < n$ and choose it maximally from among all such prime numbers. Define $\lambda_{n,p}$ to be the largest power of $\zeta_{n,p}$ which divides $p^n - 1$. We present the following special case of [2, Lemma 4.1].

Lemma 2.3 *Assume that $d \geq 9$, $G = \mathrm{SL}_d(p^a)$ and $Z \leq Z(G)$. Then*

  (i) *G/Z has a hyperbolic triple of type $(\lambda_{ad,p}, \lambda_{ad,p}, \lambda_{a(d-1),p})$; and*

 (ii) *G/Z has a hyperbolic triple of type*

$$(\lambda_{a(d-3),p}\lambda_{3a,p},\, \lambda_{a(d-3),p}\lambda_{3a,p},\, \lambda_{a(d-2),p})$$

where $\lambda_{6,2} = \lambda_{3.2,2}$ is 7.

In particular, G/Z has a Beauville structure.

Combining this with the Guralnick-Tiep Theorem we easily obtain the following theorem.

Theorem 2.4 *Suppose that p is an odd prime, $G \cong \mathrm{PSL}_d(p^a)$ and that X is a p-Frattini cover of G. If $d \geq 9$, then X has a Beauville structure.*

Proof Since p is odd and the types of the triples listed in Lemma 2.3 (i) and (ii) are coprime to $p(p^a - 1)$, Corollary 2.2 applies to give the result.   □

Suppose that X is the profinite group $\mathrm{SL}_d(\mathbb{Z}_p)$ where $\mathbb{Z}_p$ denotes the ring of p-adic integers. A calculation of Serre [10, Lemma 3 and Exercise 1a), pages IV-23 and IV-27] shows that the non-trivial finite quotients of X are p-Frattini covers whenever $p \geq 5$. Combining this with Theorem 2.4 yields the following statement.

Example 2.5 If $d \geq 9$ and $p \geq 5$, then every finite quotient of $X = \mathrm{SL}_d(\mathbb{Z}_p)$ has a Beauville structure.

3 Semidirect Products with Beauville Structures

In this section, we present some observations about Beauville structures in semidirect products $H = GV$. Lifting hyperbolic triples from G to H is not always possible. However there is an elementary criterium which at least allows us to guarantee that there are lifts of generators of G to generators of H in the case when V is a minimal normal subgroup of H which is abelian.

Lemma 3.1 *Suppose that V is an irreducible, faithful $\mathrm{GF}(p)G$-module and let $H = GV$ be the semidirect product of G and V. Assume $\mathcal{H} = (a, b, c)$ is a hyperbolic triple in G. Then*

  (i) *For all $v, w \in V$, we have $avbwc(v^{bc}w^c)^{-1} = 1$.*

  (ii) *Suppose that $X \subseteq V$ and $Y \subseteq V$ with $|X||Y| > |V||\mathrm{H}^1(G, V)|$. Then there exist $u \in X$ and $v \in Y$ such that $H = \langle au, bv \rangle$.*

  (iii) *There exist $u, v \in V$ such that $H = \langle au, bv \rangle$.*

In particular, there are $u, v \in V$ such that $(av, bw, c(v^{bc}w^c)^{-1})$ is a hyperbolic triple in H which lifts $\mathcal{H}$.

Proof We have

$$avbwc(v^{bc}w^c)^{-1} = abcv^{bc}w^c(v^{bc}w^c)^{-1}$$
$$= 1$$

and so (i) is true.

We next prove (ii). Set $\mathcal{C} = \{\langle au, bv \rangle \mid u \in X, v \in Y\}$ and let $J = \langle au, bv \rangle \in \mathcal{C}$. Since V is an irreducible $\mathrm{GF}(p)G$-module and, since

$$JV = \langle au, bv \rangle V = \langle a, b \rangle V = GV = H,$$

we infer that either J is a complement to V or $J = H$. It follows that either (ii) holds or all members of $\mathcal{C}$ are complements to V in G. Assume that the latter is the case and let $K = \langle au_1, bv_1 \rangle \in \mathcal{C}$. If $J = K$, then $(au_1)^{-1}au = u_1^{-1}u \in J \cap V$ and $v_1^{-1}v \in J \cap V$. As J is a complement to V, we deduce that $uu_1^{-1} = vv_1^{-1} = 1$. This yields $|\mathcal{C}| = |X||Y| > |V||\mathrm{H}^1(G, V)|$. However $|\mathcal{C}| \leq |V||\mathrm{H}^1(G, V)|$ which is the total number of complements to V in X. This contradiction proves (ii).

Taking $V = X = Y$, we have $|X||Y| = |V|^2 > |V||\mathrm{H}^1(G, V)|$ by the Aschbacher-Guralnick Theorem [1, Theorem 1]. Thus (iii) follows from (ii).

Lemma 3.2 *Assume that $\mathcal{B} = ((a_1, b_1, c_1), (a_2, b_2, c_2))$ is a Beauville structure of the group G. Suppose that V is an irreducible, faithful $\mathrm{GF}(p)G$-module and set $H = GV$ the semidirect product of G and V. If*

$$\dim[V, a_1] + \dim[V, b_1] > \dim V + \dim \mathrm{H}^1(G, V),$$

$$\dim[V, a_2] + \dim[V, b_2] > \dim V + \dim \mathrm{H}^1(G, V)$$

and either

(i) $C_V(c_1) = 0$ *or* $C_V(c_2) = 0$; *or*
(ii) $\langle c_1, c_2^g \rangle$ *does not centralize a vector of V for all $g \in G$,*

then H has a Beauville structure which is a lift of $\mathcal{B}$.

Proof For $i = 1, 2$, set $X_i = [V, a_i]$ and $Y_i = [V, b_i]$. Then, as $|V||\mathrm{H}^1(G, V)| < |X_i||Y_i|$, Lemma 3.1 (ii) implies that there exists $v_i \in X_i$ and $w_i \in Y_i$ such that $H = \langle a_i v_i, b_i w_i \rangle$. The choice of $v_i \in X_i$ and $w_i \in Y_i$ means that a_i and $a_i v_i$ are conjugate in H and so are b_i and $b_i w_i$. Now to form a hyperbolic triple completing $a_i v_i, b_i w_i$ we need to take a third element $d_i = c_i(v_i^{b_i c_i} w_i^{c_i})^{-1}$. Let r_i be the order of c_i, then d_i has order r_i or $p r_i$. The only way that $\mathcal{B}$ can fail to lift to a Beauville structure of H is if d_1 has order $r_1 p$ and d_2 has order $r_2 p$ and $e_1 = d_1^{r_1}$ and $e_2 = d_2^{r_2}$ are conjugate in H (and hence by elements of G). In particular, if (i) holds then $\mathcal{B}$ lifts to a Beauville structure of H. So consider the alternative case. Then there exists $g \in G$ such that $e_1 = e_2^g$ is centralized by $\langle d_1, d_2^g \rangle$. But $\langle c_1, c_2^g \rangle$ does not centralize any vector in V. This shows that $\mathcal{B}$ lifts to a Beauville structure of H.

We state the famous theorem of Scott as it shows that the initial restriction on the sizes of the commutators in Lemma 3.2 is not a severe limitation. We present the theorem just for hyperbolic triples.

Theorem 3.3 (L. Scott) *Suppose that (x_1, x_2, x_3) is a hyperbolic triple in G and let V be an irreducible G-module of dimension n. Assume that, for $i = 1, 2, 3$, $\dim[V, x_i] = d_i$. Then the following hold:*

(i) $d_1 + d_2 + d_3 \geq \dim \mathrm{H}^1(G, V) + 2n$; *and*
(ii) *if x_i each have order coprime to p and the Schur multiplier of G has order coprime to p, then $d_1 + d_2 + d_3 \geq \dim \mathrm{H}^1(G, V) + \dim \mathrm{H}^1(G, V^*) + 2n$.*

Proof This is a combination of Scott's results [11] Theorem 1 and Proposition 1 (a) for (i) and (b) for (ii) stated for the special case of hyperbolic triples. $\qquad\square$

Now we give an example which shows that hyperbolic triples do not always lift so that the elements in the triple have the same order.

Example 3.4 Suppose that $G \cong \mathrm{SL}_3(2)$ and V is the natural 3-dimension G-module. Then G has a hyperbolic triple (a, b, c) with $o(a) = 2$, $o(b) = 3$ and $o(c) = 7$. We know $|[V, a]| = 2$, $|[V, b]| = 2^2$ and $V = [V, c]$. It is easy to check that GV has no hyperbolic triple (a', b', c') such that $o(a') = 2$, $o(b') = 3$, $o(c') = 7$.

Theorem 3.5 *Suppose that $G \cong \mathrm{SL}_d(p^a)$ with $d \geq 9$ and V is the natural $\mathrm{GF}(p^a)G$-module. Then $H = GV$ has a Beauville structure.*

Proof By Lemma 2.3, G has a hyperbolic triple of type $(\lambda_{ad,p}, \lambda_{ad,p}, \lambda_{a(d-1),p})$ and a hyperbolic triple of type $(\lambda_{a(d-3),p}\lambda_{3a,p}, \lambda_{a(d-3),p}\lambda_{3a,p}, \lambda_{a(d-2),p})$ where $\lambda_{6,2} = \lambda_{3,2,2}$ is 7. The elements of order $\lambda_{ad,p}$ and $\lambda_{a(d-3),p}\lambda_{3a,p}$ act fixed point freely on V and the elements of order $\lambda_{a(d-1),p}$ and $\lambda_{a(d-2),p}$ fix a 1 and a 2-space in V respectively. Furthermore, by [8, Table B], $H^1(G, V) = 0$ and so $H = GV$ has a Beauville structure by Lemma 3.2. ∎

4 Questions

In this section we pose some questions which we hope suggest fruitful research directions for the study of hyperbolic triples and Beauville surfaces. For Frattini covers we ask

Question 4.1 *For some of the small rank quasisimple classical groups G in characteristic p, one of the hyperbolic triples describing the Beauville structure given in [2] involves a p-element. This means that we may not apply Theorem 2.1 to lift the hyperbolic triple to a p-Frattini cover of G. Under what hypotheses can a hyperbolic triple be lifted to a hyperbolic triple for G?*

Since the Guralnick-Tiep Theorem does not apply when $p = 2$, it would be nice to know the answer to the following question.

Question 4.2 *Under which additional hypotheses can the Guralnick-Tiep Theorem be extended to 2-Frattini covers?*

The next question, which we assume typically has an affirmative answer, is a special case of the last question.

Question 4.3 *The groups $2_-^{1+2n}.\Omega_{2n}^-(2)$, $2_-^{1+2n}.\Omega_{2n}^-(2)$ and $\mathbb{Z}/4\mathbb{Z} \circ 2_+^{1+2n}.\mathrm{Sp}_{2n}(2)$ are derived subgroups of groups in the Aschbacher class $\mathcal{C}_6$ of maximal subgroups of classical groups. For sufficiently large n, they are 2-Frattini covers of their simple quotients. Do they have Beauville structures?*

The natural questions for semidirect products seem to be

Question 4.4 *For a quasisimple group G and prime p, for which irreducible $\mathrm{GF}(p)G$-modules is the semidirect product GV a Beauville group?*

In [6, 7] Gareth Jones begins the investigation of Beauville structures in direct products H^k of k non-abelian simple groups isomorphic to H (characteristically simple groups). He conjectures that such groups have Beauville structures whenever they can be generated by two elements (other than $H = \mathrm{Alt}(5)$ and $k = 1$). He proves his conjecture for $\mathrm{Alt}(n)$, $n \geq 5$, $\mathrm{PSL}_2(q)$, $q \geq 4$, $\mathrm{PSL}_3(q)$, $\mathrm{PSU}_3(q)$, $q > 2$, $^2B_2(2^e)$, $e > 1$ odd, $^2G_2(3^e)$, $e > 1$ odd, and for the sporadic simple groups. We can pose a similar question for semidirect products which extends Question 4.4.

Question 4.5 *Suppose that G is a quasisimple group, p is a prime and V is an irreducible* $\mathrm{GF}(p)G$*-module. Let nV denote a direct sum of n copies of V. For which n is $G(nV)$ a Beauville group?*

The next example, though not using a hyperbolic triple, shows that we may expect that an answer to Question 4.5 which is similar to that conjectured in [6, 7] for characteristically simple non-abelian groups.

Example 4.6 Let $G = \mathbb{Z}/3\mathbb{Z}$ and V be elementary abelian of order 4 admitting G non-trivially. Then $GV \cong \mathrm{Alt}(4)$ and G has a triple of type $(3, 3, 3)$ (which is not hyperbolic). The universal $(3, 3, 3)$-group is isomorphic to $\mathbb{Z}^2$ extended by a cyclic group of order 3. Now, in $G(nV)$ we see that every element has order either 2 or 3. If a $(3, 3, 3)$ triple for GV could be extended to a $(3, 3, 3)$-triple for $G(nV)$, then that triple would also be a $(3, 3, 3)$-triple. Hence it must generate a subgroup of $G(nV)$ isomorphic to $\mathrm{Alt}(4)$. We conclude that $n = 1$.

Based on our results in the previous section, our most general question involves one-headed groups. Recall that a group X is *one-headed* with *head G* provided it has a unique maximal normal subgroup Y and $X/Y \cong G$ is a non-abelian simple group. Clearly the quasisimple groups, Frattini covers and the semidirect products in Lemma 3.2 are one-headed.

Question 4.7 *Let X be a one-headed group with head G. Under which hypotheses do Beauville structures of G lift to Beauville structures of X?*

Magaard's talk at the meeting in Newcastle raised questions about the number of Beauville structures that a given quasisimple group might admit. We close this article by stating this question explicitly and then setting it in a more specific context.

Question 4.8 (Magaard, 2012) *Given a quasisimple group G, how many Beauville structures does G admit?*

Question 4.9 *Given a non-abelian simple group G, a Beauville structure $\mathcal{B}$ of G and an extension X of G, how many Beauville structures of X project onto $\mathcal{B}$?*

Question 4.10 *Given a group extension X with $X/Y \cong G$ a simple group and a hyperbolic triple $\mathcal{H}$ of G, how many hyperbolic triples project on to $\mathcal{H}$?*

We expect the last three questions may have a probabilistic answer rather than an explicit integer value.

Of course, ultimately these results are closely connected to the 1- and 2-cohomology of G. The 3-cohomology of groups may be needed to investigate Beauville structures of one-headed non-abelian extensions.

References

1. M. Aschbacher, R.M. Guralnick, Some applications of the first cohomology group. J. Algebra **90**(2), 446–460 (1984)
2. B. Fairbairn, K. Magaard, C. Parker, Generation of finite quasisimple groups with an application to groups acting on Beauville surfaces. Proc. LMS appeared online 18 February, (2013)
3. W. Feit, On large Zsigmondy primes. Proc. Am. Math. Soc. **102**(1), 29–36 (1988)
4. R. Guralnick, G. Malle, Simple groups admit Beauville structures. J. Lond. Math. Soc. **85**(3), 694–721 (2012)
5. R.M. Guralnick, P.H. Tiep, Lifting in Frattini covers and a characterization of finite solvable groups. arXiv:1112.4559
6. G.A. Jones, Characteristically simple Beauville groups, I: Cartesian powers of alternating groups. arXiv:1304.5444
7. G.A. Jones, Characteristically simple Beauville groups, II: low rank and sporadic groups. arXiv:1304.5450
8. W. Jones, B. Parshall, On the 1-cohomology of finite groups of Lie type, in *Proceedings of the Conference on Finite Groups* (Univ. Utah, Park City, Utah, 1975)
9. Darren Semmen, The group theory behind modular towers. Séminaires Congrès **13**, 343–366 (2006)
10. J.-P. Serre, Abelian l-adic representations and elliptic curves. McGill University lecture notes written with the collaboration of Willem Kuyk and John Labute, W. A. Benjamin, Inc., New York-Amsterdam (1968)
11. L.L. Scott, Matrices and cohomology, Ann. Math. (2) 105 (1977), no. 3, 473–492

Surfaces Isogenous to a Product of Curves, Braid Groups and Mapping Class Groups

Matteo Penegini

Abstract We present some group theoretical methods to give bounds on the number of connected components of the moduli space of surfaces of general type, focusing on some families of regular surfaces isogenous to a product of curves.

1 Introduction

This article is a revised version of the talk I gave at the conference "*Beauville Surfaces and groups*" held in Newcastle in June 2012. It presents some group theoretical methods to give bounds on the number of connected components of the moduli space of surfaces of general type, focusing on some families of regular surfaces isogenous to a product of curves. Some of the results appearing in this work have been proven in collaboration with Shelly Garion.

We will use the standard notation from the theory of complex algebraic surfaces. Let S be a smooth, complex, projective, *minimal* surface S of *general type*; this means that the canonical divisor K_S of S is *big* and *nef*.

The principal numerical invariants for the study of minimal surfaces of general type are

- the *geometric genus* $p_g(S) := h^0(S, \Omega_S^2) = h^0(S, \mathcal{O}_S(K_S))$,
- the *irregularity* $q(S) := h^0(S, \Omega_S^1)$, and
- the *self intersection of the canonical divisor* K_S^2.

In fact, these determine all the other classical invariants, as

- the *Euler-Poincaré characteristic* $\chi(\mathcal{O}_S) = 1 - q(S) + p_g(S)$,
- the *topological Euler number* $e(S) = 12\chi(S) - K_S^2$, and
- the *plurigenera* $P_n(S) = \chi(S) + \binom{n}{2}K_S^2$.

M. Penegini (✉)
Matteo Penegini, Dipartimento di Matematica "Federigo Enriques", Università degli Studi di Milano, Via Saldini 50, 20133 Milano, Italy
e-mail: matteo.penegini@unimi.it

© Springer International Publishing Switzerland 2015
I. Bauer et al. (eds.), *Beauville Surfaces and Groups*, Springer Proceedings
in Mathematics & Statistics 123, DOI 10.1007/978-3-319-13862-6_9

Moreover, we call a surface *regular* if its irregularity vanishes, i.e., $q(S) = 0$.

By a theorem of Bombieri, a minimal surface of general type S with fixed invariants is birationally mapped to a normal surface X in a fixed projective space of dimension $P_5(S) - 1$. Moreover, X is uniquely determined and is called the *canonical model* of S. Let us recall Gieseker's Theorem.

Theorem 1.1 *There exists a quasi-projective coarse moduli space $\mathcal{M}_{y,x}$ for canonical models of surfaces of general type S with fixed invariants $y := K_S^2$ and $x := \chi$.*

The union $\mathcal{M}$ over all admissible pairs of invariants (y, x) of these spaces is called the *moduli space of surfaces of general type*. If S is a smooth minimal surface of general type, we denote by $\mathcal{M}(S)$ the subvariety of $\mathcal{M}_{y,x}$, corresponding to surfaces (orientedly) homeomorphic to S. Moreover, we denote by $\mathcal{M}_{y,x}^0$ the subspace of the moduli space corresponding to regular surfaces.

It is known that the number of connected components $\delta(y, x)$ of $\mathcal{M}_{y,x}^0$ is bounded from above by a function of y; more precisely by [7] we have $\delta(y, x) \leq cy^{77y^2}$, where c is a positive constant. Hence we have that the number of components has an exponential upper bound in K^2.

There are also some results regarding the lower bound. In [21], for example, Manetti constructed a sequence S_n of simply connected surfaces of general type, such that the lower bound for the number of the connected components $\delta(S_n)$ of $\mathcal{M}(S_n)$ is given by

$$\delta(S_n) \geq y_n^{\frac{1}{5} log y_n}.$$

Using group theoretical methods we are able to describe the asymptotic growth of the number of connected components of the moduli space of surfaces of general type relative to certain sequences of surfaces. More precisely, we apply the definition and some properties of regular surfaces isogenous to a product of curves and of some special cases of them, Beauville surfaces, to reduce the geometric problem of finding connected components into the algebraic one of counting orbits of some group action, which can be effectively computed.

The paper is organized as follows.

In the first Section we recall the definition and some properties of the mapping class group. Moreover, we describe the Hurwitz moves in the most general setting.

In the second Section we briefly recall the definition of surfaces isogenous to a product of curves. We will recall some of the properties of these surfaces focusing on their moduli space.

In the third part we present some results obtained with Shelly Garion about the number of connected components of the moduli space of surfaces isogenous to a product.

Finally, we also point out some possible future developments.

2 Braid Group and Mapping Class Group

In this section, we first recall the definition of mapping class group. Next we give a presentation of it for $\mathbb{P}^1 - \{p_1, \ldots, p_r\}$, and more generally for a curve of genus g' with r marked points. After that, we calculate the Hurwitz moves induced by those groups. We mainly follow the definitions and notation of [9].

Definition 2.1 Let M be a differentiable manifold, then the *mapping class group* of M is the group:

$$\mathrm{Map}(M) := \pi_0(\mathrm{Diff}^+(M)) := \mathrm{Diff}^+(M)/\mathrm{Diff}^0(M),$$

where $\mathrm{Diff}^+(M)$ is the group of orientation preserving diffeomorphisms of M and $\mathrm{Diff}^0(M)$ is the subgroup of diffeomorphisms of M isotopic to the identity.

If M is a compact complex curve of genus g' we use the following notation:

(1) We denote the mapping class group of M without marked points by $\mathrm{Map}_{g'}$.
(2) If we consider r unordered marked points $p_1, \ldots, p_r$ on M we define:

$$\mathrm{Map}_{g',[r]} = \pi_0(\mathrm{Diff}^+(M - \{p_1, \ldots, p_r\})),$$

and this is known as the *full mapping class group*.

There is a way to present the full mapping class group of a curve using three different types of twists.

Theorem 2.2 *The mapping class group* $\mathrm{Map}_{0,[r]} = \pi_0(\mathrm{Diff}^+(\mathbb{P}^1 - \{p_1, \ldots, p_r\}))$ *is isomorphic to the* braid group $\mathbf{B}_r$ *on r strands, which can be presented as*

$$\mathbf{B}_r = \langle \sigma_1, \ldots, \sigma_{r-1} | \sigma_i\sigma_{i+1}\sigma_i = \sigma_{i+1}\sigma_i\sigma_{i+1}, \sigma_i\sigma_j = \sigma_j\sigma_i \text{ if } |i - j| \geq 2 \rangle.$$

For a proof of the above Theorem see, for example, [6, Theorem 1.11].

In this way Artin's standard generators σ_i $(i = 1, \ldots, r - 1)$ of $\mathbf{B}_r$ can be represented by the so-called half-twists.

Definition 2.3 The *half-twist* σ_j is a diffeomorphism of $\mathbb{P}^1 - \{p_1, \ldots, p_r\}$ isotopic to the homeomorphism given by (see Fig. 1):

- A rotation of $180°$ on the disk with center $j + \frac{1}{2}$ and radius $\frac{1}{2}$;
- on a circle with the same center and radius $\frac{2+t}{4}$ the map σ_j is the identity if $t \geq 1$ and a rotation of $180(1 - t)°$, if $t \leq 1$.

We want to give a similar presentation for a group $\mathrm{Map}_{g'}$ with $g' \geq 1$, so we have to introduce the Dehn twists.

Definition 2.4 Let C be an oriented Riemann surface. Then a *positive Dehn twist* t_α with respect to a simple closed curve α on C is an isotopy class of a diffeomorphism

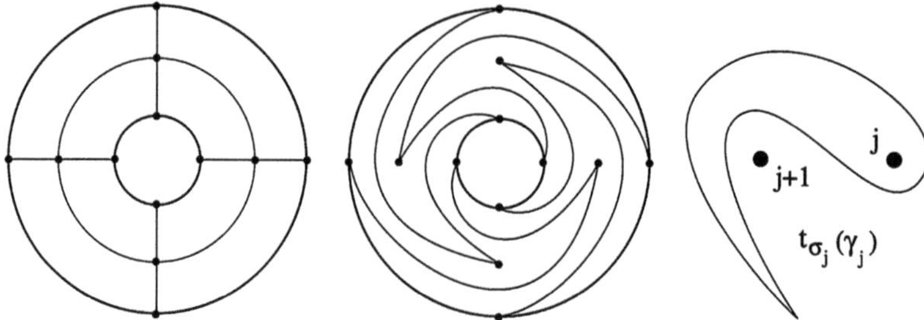

Fig. 1 The half–twist

h of C which is equal to the identity outside a neighborhood of α orientedly homeomorphic to an annulus in the plane, while inside the annulus h rotates the inner boundary of the annulus by $360°$ to the right and damps the rotation down to the identity at the outer boundary (see Fig. 2).

We have then the following classical results of Dehn [13].

Theorem 2.5 *The mapping class group* $\mathrm{Map}_{g'}$ *is generated by Dehn twists.*

We give the generators of the group $\mathrm{Map}_{g'}$.

Theorem 2.6 *The group* $\mathrm{Map}_{g'}$ *is generated by the Dehn twists with respect to the curves in the Fig. 3.*

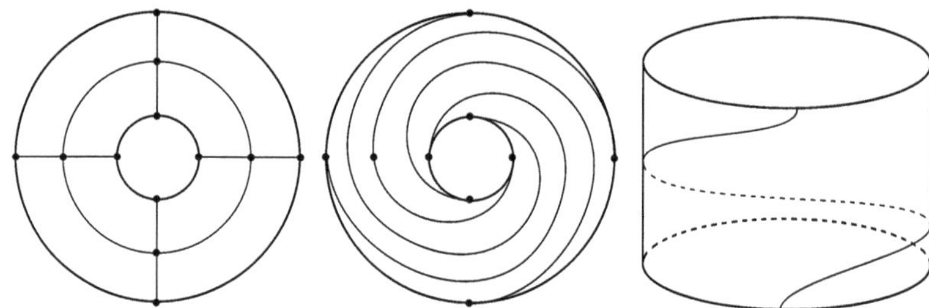

Fig. 2 The positive Dehn twist

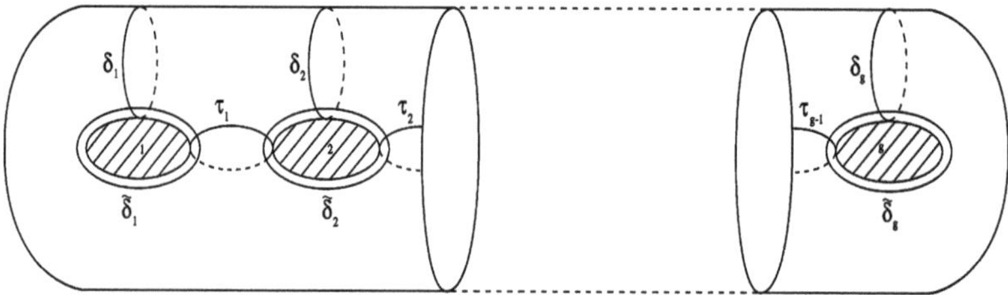

Fig. 3 Generators for $\mathrm{Map}_{g'}$

A proof of the above Theorem can be found in [6, Theorem 4.8].

For a more general situation where the surface has $g' > 0$ and r marked points we need to introduce a third type of twist: The ξ-twists, which link the holes with the marked points. Let us recall the Birman short exact sequence for a Riemann surface C' with $g(C') > 1$ or $r > 1$, setting $\mathcal{B} := \{p_1, \ldots, p_r\}$:

$$1 \longrightarrow \pi_1(C' \setminus \mathcal{B}, p) \xrightarrow{\Xi} \pi_0(\text{Diff}^+(C'\setminus(\mathcal{B} \cup \{p\}))) \longrightarrow \pi_0(\text{Diff}^+(C'\setminus\mathcal{B})) \longrightarrow 1. \tag{1}$$

The map Ξ can be described as follows (cf. [5]). Let $[\gamma] \in \pi_1(C' \setminus \mathcal{B}, p)$ and γ be a simple, smooth loop based at p representing $[\gamma]$. Then $\Xi([\gamma])$ is the isotopy class of a ξ-twist with respect to the closed curve γ. In addition, this new twist is isotopic to the identity in $\pi_0(\text{Diff}^+(C \setminus (\mathcal{B}))$. Let us now describe a ξ-twist. We shall consider the annulus $A := \{z = \rho e^{i\theta} \in \mathbb{C} | 1 \leq \rho \leq 2\}$, and we define $h : A \to A$ as follows

$$h(\rho, \theta) : \begin{cases} \left(\rho, \theta - 4\pi(\rho - 1)\right) & 1 \leq \rho \leq \frac{3}{2} \\ \left(\rho, \theta - 4\pi(2 - \rho)\right) & \frac{3}{2} \leq \rho \leq 2. \end{cases} \tag{2}$$

Definition 2.7 Let C be a Riemann surface, and α a simple closed curve on C. Let ι be a diffeomorphism between A and a tubular neighborhood of α. Then the ξ-*twist* $\mathbf{t}_\alpha$ with respect to α is defined as $\iota \circ h \circ \iota^{-1} |_{\iota(A)}$ extended to the whole C as the identity on $C \setminus \iota(A)$ (see Fig. 4).

Therefore, for the more general situation we have the following.

Theorem 2.8 ([5, Theorem 3]) *Let* $g(C') \neq 0$ *and* $g(C') > 1$ *or* $r > 1$ *then the group* $\text{Map}_{g',[r]}$ *is generated by the* $3g' - 1$ *Dehn twists with respect to the curves* δ_j, $\tilde{\delta}_j$ *and* τ_j, *by the* $2rg'$ ξ-*twists with respect to the curves* $\xi^l_{j,d}$ *and the* $r - 1$ *half-twists about the points* $p_1, \ldots, p_r$ *in Fig. 5.*

Let C' be a Riemann surfaces of genus g', and let $\mathcal{B} := \{p_1, \ldots, p_r\}$ a set of points on C'. A *geometric basis* of $\pi_1(C' \setminus \mathcal{B}, p_0)$ consists of simple non-intersecting (away

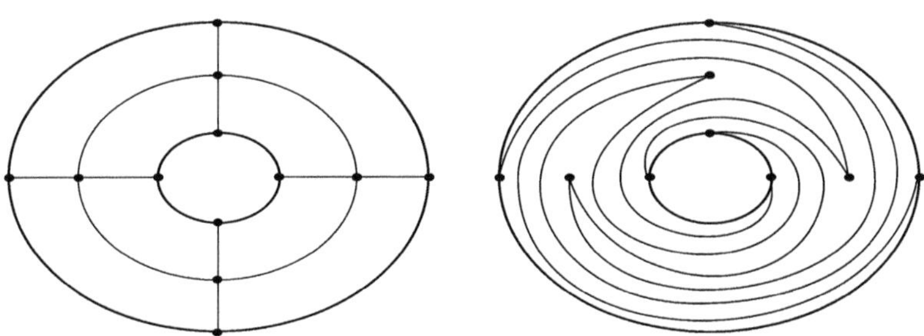

Fig. 4 The ξ–twist

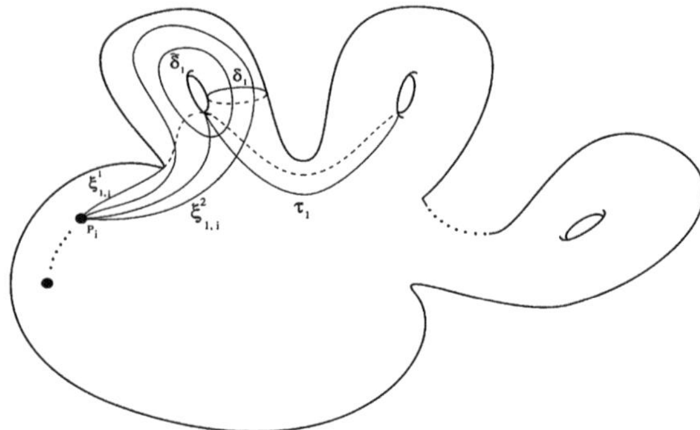

Fig. 5 Generators of $\mathrm{Map}_{g',[r]}$

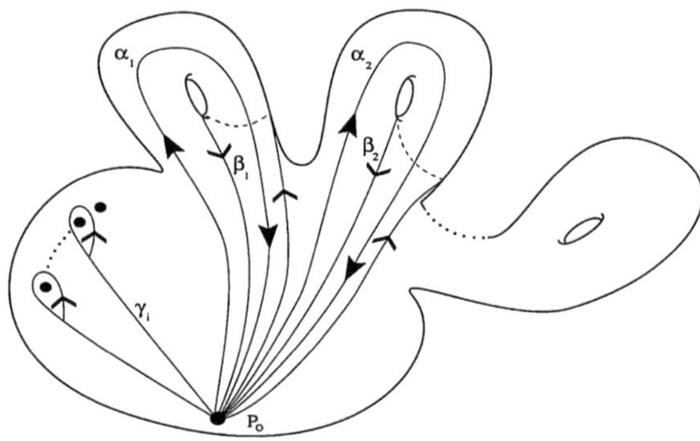

Fig. 6 Geometric basis of $\pi_1(C' \setminus \mathcal{B}, p_0)$

from the base point) loops (see Fig. 6)

$$\gamma_1, \ldots, \gamma_d, \alpha_1, \beta_1, \ldots, \alpha_{g'}, \beta_{g'}$$

such that we get the presentation

$$\pi_1(C \setminus \mathcal{B}, p_0) := \left\langle \alpha_1, \beta_1, \ldots, \alpha_{g'}, \beta_{g'}, \gamma_1, \ldots, \gamma_r \,\middle|\, \gamma_1 \cdot \ldots \cdot \gamma_r \cdot \prod_{k=1}^{g'} [\alpha_k, \beta_k] \right\rangle.$$

Definition 2.9 Let $g', m_1, \ldots, m_r$ be positive integers. An *orbifold surface group* of type $(g' \mid m_1, \ldots, m_r)$ is a group presented as follows:

$$\Gamma(g' \mid m_1, \ldots, m_r) := \left\langle a_1, b_1, \ldots, a_{g'}, b_{g'}, c_1, \ldots, c_r \,\middle|\, \right.$$

$$\left. c_1^{m_1} = \cdots = c_r^{m_r} = c_1 \cdot \ldots \cdot c_r \cdot \prod_{k=1}^{g'} [a_k, b_k] = 1 \right\rangle.$$

We notice that the choice of a geometric basis yields an obvious epimorphism
$\pi_1(C \setminus \mathcal{B}, p_0) \to \Gamma(g' \mid m_1, \ldots, m_r)$.

The following is a reformulation of *Riemann's existence theorem*:

Theorem 2.10 *A finite group G acts as a group of automorphisms on some compact Riemann surface C of genus g if and only if there are natural numbers $g', m_1, \ldots, m_r,$ and an orbifold homomorphism*

$$\theta: \Gamma(g' \mid m_1, \ldots, m_r) \to G \tag{3}$$

such that $ord(\theta(c_i)) = m_i$ for all i and such that the Riemann-Hurwitz relation holds:

$$2g - 2 = |G| \left(2g' - 2 + \sum_{i=1}^{r} \left(1 - \frac{1}{m_i} \right) \right). \tag{4}$$

If this is the case, then g' is the genus of $C' := C/G$. The G-cover $C \to C'$ is branched at r points $p_1, \ldots, p_r$ with branching indices $m_1, \ldots, m_r$, respectively. Let $\Gamma = \Gamma(g' \mid m_1, ..., m_r)$ be an orbifold surface group with a presentation as in Definition 2.9. If G is a finite group quotient of Γ as in (3), then we say that G is $(g' \mid m_1, \ldots, m_r)$−generated, the image of the generators of Γ in G is called a *system of generators* for G. Finally, θ is called an *admissible epimorphism*.

Definition 2.11 An automorphism $\eta \in \mathrm{Aut}(\Gamma)$ is said to be *orientation preserving* if the action induced on $\langle \alpha_1, \beta_1, \ldots, \alpha_{g'}, \beta_{g'} \rangle^{ab}$ has determinant $+1$, and for all $i \in \{1, \ldots, r\}$ there exists j such that $\eta(\gamma_i)$ is conjugate to γ_j, which implies $ord(\gamma_i) = ord(\gamma_j)$.

The subgroup of orientation preserving automorphisms of Γ is denoted by $\mathrm{Aut}^+(\Gamma)$ and the quotient $\mathrm{Out}^+(\Gamma) := \mathrm{Aut}^+(\Gamma)/\mathrm{Inn}(\Gamma)$ is called the *mapping class group* of Γ.

Theorem 2.12 *Let $\Gamma = \Gamma(g' \mid m_1, \ldots, m_r)$ be an orbifold surface group. Then there is an isomorphism of groups:*

$$\mathrm{Out}^+(\Gamma) \cong \mathrm{Map}_{g',[r]}.$$

This is a classical result cf. e.g., [20, Sect. 4] .

Moreover let G be a finite group $(g' \mid m_1, \ldots, m_r)$—generated. There is a section $s : \mathrm{Out}^+(\Gamma) \to \mathrm{Aut}^+(\Gamma)$, which induces an action of the $\mathrm{Map}_{g',[r]}$ on the generators of Γ. Such action does not depend on s up to simultaneous conjugation, meaning that the action is defined up to inner automorphisms. This action induces an action on the systems of generators of G via composition with admissible epimorphisms.

Definition 2.13 Let G be a finite group $(g' \mid m_1, \ldots, m_r)$—generated. If two systems of generators $\mathcal{V}_1$ and $\mathcal{V}_2$ are in the same $\mathrm{Map}_{g',[r]}$-orbit, we say that they are related by a *Hurwitz move* (or are *Hurwitz equivalent*).

Proposition 2.14 *Let C' be a curve of genus g', $\mathcal{B} := \{p_1, \ldots, p_r\}$, and with $g' \neq 0$ and $g' > 1$ or $r > 1$. Up to inner automorphisms, the action of $\mathrm{Map}_{g',[r]}$ on $\Gamma(g' \mid m_1 \ldots m_r)$ is induced by the following action on a geometric basis of $\pi_1(C' \setminus \mathcal{B}, p_0)$*

$$
\mathbf{t_j} : \begin{cases} \alpha_j \mapsto \alpha_j \beta_j^{-1} \\ \alpha_i \mapsto \alpha_i & \forall i \neq j \\ \beta_i \mapsto \beta_i & \forall i \\ \gamma_i \mapsto \gamma_i & \forall i \end{cases}
\qquad
\mathbf{t_{\tilde{\delta}_j}} : \begin{cases} \alpha_i \mapsto \alpha_i & \forall i \\ \beta_j \mapsto \beta_j \alpha_j \\ \beta_i \mapsto \beta_i & \forall i \neq j \\ \gamma_i \mapsto \gamma_i & \forall i \end{cases}
$$

$$
\mathbf{t_{\sigma_h}} : \begin{cases} \alpha_i \mapsto \alpha_i & \forall i \\ \beta_i \mapsto \beta_i & \forall i \\ \gamma_h \mapsto \gamma_{h+1} \\ \gamma_{h+1} \mapsto \gamma_{h+1}^{-1} \gamma_h \gamma_{h+1} \\ \gamma_i \mapsto \gamma_i & \forall i \neq h, h+1 \end{cases}
\qquad
\mathbf{t_{\tau_k}} : \begin{cases} \alpha_k \mapsto \alpha_k \eta_k^{-1} \\ \beta_k \mapsto \beta_k^{\eta_k} \\ \alpha_{k+1} \mapsto \eta_k \alpha_k \\ \alpha_i \mapsto \alpha_i & \forall i \neq k, k+1 \\ \beta_i \mapsto \beta_i & \forall i \neq k \\ \gamma_i \mapsto \gamma_i & \forall i \end{cases}
$$

$$
\mathbf{t_{\xi_{j,d}^1}} : \begin{cases} \alpha_j \mapsto \chi_{j,d} \alpha_j \\ \alpha_i \mapsto \alpha_i & \forall i \neq j \\ \beta_i \mapsto \beta_i & \forall i \\ \gamma_d \mapsto \gamma_d^{\epsilon_{j,d}} \\ \gamma_i \mapsto \gamma_i & \forall j \neq d \end{cases}
\qquad
\mathbf{t_{\xi_{j,d}^2}} : \begin{cases} \alpha_i \mapsto \alpha_i & \forall i \\ \beta_j \mapsto \alpha_j^{-1} \chi_{j,d} \alpha_j \beta_j \\ \beta_i \mapsto \beta_i & \forall i \neq j \\ \gamma_d \mapsto \gamma_d^{\epsilon'_{j,d}} \\ \gamma_i \mapsto \gamma_i & \forall j \neq d \end{cases}
$$

for $1 \leq j \leq g'$, $1 \leq k \leq (g'-1)$, $1 \leq h \leq (r-1)$, and $1 \leq d \leq r$. Moreover we set $\eta_k := \beta_k^{-1} \alpha_{k+1} \beta_{k+1} \alpha_{k+1}^{-1}$, $\chi_{j,d} := (\Pi_{k=1}^{j-1}[\alpha_k, \beta_k])^{-1} \gamma_d \Pi_{k=1}^{j-1}[\alpha_k, \beta_k]$, $\epsilon_{j,d} := \gamma_d(\Pi_{k=1}^{j}[\alpha_k, \beta_k])\alpha_j \beta_j \alpha_j^{-1}(\Pi_{k=1}^{j}[\alpha_k, \beta_k])^{-1}$, and $\epsilon'_{j,d} := \gamma_d(\Pi_{k=1}^{j}[\alpha_k, \beta_k])\alpha_j^{-1}(\Pi_{k=1}^{j}[\alpha_k, \beta_k])^{-1}$.

In the above proposition the twists $\mathbf{t_{\tilde{\delta}_j}}$, $\mathbf{t_{\delta_j}}$ and $\mathbf{t_{\tau_j}}$ correspond to Dehn twists, $\mathbf{t_{\xi_{r_{j,d}}}}$ and $\mathbf{t_{\xi_{r_{j,d}}}}$ to ξ-twists, and finally $\mathbf{t_{\sigma_h}}$ to half-twists.

Proof One notices that a Riemann surface of genus g' is a connected sum of g' tori. Then one can use the Fig. 7 to calculate the Dehn twists about the curves δ_j, and similarly for the Dehn twists about the curves $\tilde{\delta}_j$. One can use the results given in [22] to calculate the Dehn twists about the curves τ_j. In the Appendix of [12] are described the actions of the ξ-twists. Finally the half-twists action is clear by Fig. 1. $\qquad\square$

For the case $g(C') = 1$ and $r = 1$ see e.g., [24] for a proof of the following proposition.

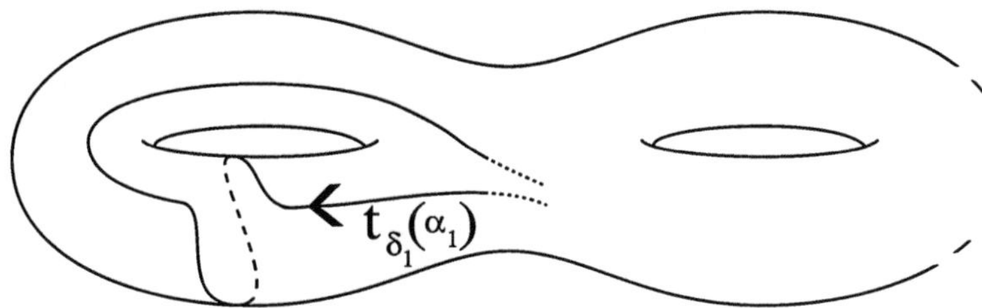

Fig. 7 Example $t_{\delta_1}(\alpha_1)$

Proposition 2.15 ([24, Proposition 1.10]) *Up to inner automorphisms, the action of* $\mathrm{Map}_{1,1}$ *on* $\Gamma(1 \mid m_1)$ *is induced by the following action on a geometric basis of* $\pi_1(C' \setminus \mathcal{B}, p_0)$

$$
\mathbf{t}_{\tilde{\delta}} : \begin{cases} \alpha_1 \to \alpha_1 \\ \beta_1 \to \beta_1 \alpha_1 \\ \gamma_1 \to \gamma_1 \end{cases}
\quad
\mathbf{t}_{\delta} : \begin{cases} \alpha_1 \to \alpha_1 \beta_1^{-1} \\ \beta_1 \to \beta_1 \\ \gamma_1 \to \gamma_1. \end{cases}
$$

Finally we have.

Proposition 2.16 *Up to inner automorphism, the action of* $\mathrm{Map}_{0,[r]}$ *on* $\Gamma(0 \mid m_1, \ldots, m_r)$ *is induced by the following action on a geometric basis of* $\pi_1(\mathbb{P}^1 \setminus \mathcal{B}, p_0)$

$$
\sigma_i : \begin{cases} \gamma_i \to \gamma_{i+1} \\ \gamma_{i+1} \to \gamma_{i+1}^{-1} \gamma_i \gamma_{i+1} \\ \gamma_j \to \gamma_j \ \text{if } j \neq i, i+1, \end{cases}
$$

for $i = 1, \ldots, r-1$.

3 Surfaces Isogenous to a Product of Curves

A surface S is said to be *isogenous to a (higher) product of curves* if and only if S is a quotient $S := (C_1 \times C_2)/G$, where C_1 and C_2 are curves of genus at least two, and G is a finite group acting freely on $C_1 \times C_2$.

Let S be a surface isogenous to a higher product, and $G^\circ := G \cap (Aut(C_1) \times Aut(C_2))$. Then G° acts on the two factors C_1 and C_2 and diagonally on the product $C_1 \times C_2$. If G° acts faithfully on both curves, we say that $S = (C_1 \times C_2)/G$ is a *minimal realization* of S. In [8], the author proves that any surface isogenous to a higher product admits a unique minimal realization. From now on, we work only with minimal realizations.

There are two cases: the *mixed* case where the action of G exchanges the two factors (in this case C_1 and C_2 are isomorphic and $G^\circ \neq G$); the *unmixed* case (where $G = G^\circ$, and therefore it acts diagonally).

Moreover, we observe that a surface isogenous to a product of curves is of general type. It is always minimal and its numerical invariants are explicitly given in terms of the genera of the curves and the order of the group. Indeed, we have the following proposition.

Proposition 3.1 *Let $S = (C_1 \times C_2)/G$ be a surface isogenous to a higher product of curves, then:*

$$\chi(S) = \frac{(g(C_1) - 1)(g(C_2) - 1)}{|G|}, \quad e(S) = 4\chi(S), \quad K_S^2 = 8\chi(S). \qquad (5)$$

The irregularity of these surfaces is easily computed by

$$q(S) = g(C_1/G) + g(C_2/G). \qquad (6)$$

By the above formula a surface S isogenous to a product of curves has $q(S) = 0$ if and only if the two quotients C_i/G are isomorphic to $\mathbb{P}^1$. Moreover, if both coverings $C_i \to C_i/G \cong \mathbb{P}^1$ are ramified in exactly 3 points, S is a *Beauville surface*. This last condition is equivalent to saying that Beauville surfaces are rigid, i.e., have no nontrivial deformations.

In the unmixed case G acts separately on C_1 and C_2, and the two projections $\pi_i : C_1 \times C_2 \longrightarrow C_i$ for $i = 1, 2$ induce two isotrivial fibrations $\alpha_i : S \longrightarrow C_i/G$ for $i = 1, 2$, whose smooth fibres are isomorphic to C_2 and C_1, respectively. *We work only with surfaces of unmixed type.*

Working out the definition of surfaces isogenous to a product, one sees that there is a pure group theoretical condition which characterizes the groups of such surfaces: the existence of a *"ramification structure"*.

Definition 3.2 Let G be a finite group and $\theta_1 : \Gamma(g_1' \mid m_{1,1}, \ldots, m_{1,r_1}) \twoheadrightarrow G$ an admissible epimorphism. Let $\mathcal{V}_1$ be the system of generators of G induced by θ_1, i.e., the elements of G which are images of the generators of Γ. We say that $\mathcal{V}_1$ is of type $\tau_1 := (g_1' \mid m_{1,1}, \ldots, m_{1,r_1})$.

Moreover, let $\theta_2 : \Gamma(g_2' \mid m_{2,1}, \ldots, m_{2,r_2}) \twoheadrightarrow G$ be another admissible epimorphism and $\mathcal{V}_2$ be the system of generators of G induced by θ_2. Then $\mathcal{V}_1$ and $\mathcal{V}_2$ are said to be *disjoint*, if:

$$\Sigma(\mathcal{V}_1) \bigcap \Sigma(\mathcal{V}_2) = \{1\}, \qquad (7)$$

where

$$\Sigma(\mathcal{V}_i) := \bigcup_{g \in G} \bigcup_{j=0}^{\infty} \bigcup_{k=1}^{r_i} g \cdot \theta_i(\gamma_k)^j \cdot g^{-1}.$$

Definition 3.3 Let $\tau_i := (g_i' \mid m_{1,i}, \ldots, m_{r_i,i})$ for $i = 1, 2$ be two types. An *unmixed ramification structure* of type (τ_1, τ_2) for a finite group G, is a pair $(\mathcal{V}_1, \mathcal{V}_2)$ of disjoint systems of generators of G, whose types are τ_i, and they satisfy:

$$\mathbb{Z} \ni \frac{|G| \left(2g_i' - 2 + \sum_{l=1}^{r_i} \left(1 - \frac{1}{m_{i,l}} \right) \right)}{2} + 1 \geq 2, \tag{8}$$

for $i = 1, 2$.

We shall denote by $\mathcal{U}(G; \tau_1, \tau_2)$ the set of all pairs $(\mathcal{V}_1, \mathcal{V}_2)$ of disjoint systems of generators of unordered type (τ_1, τ_2). Here *unordered type* τ means that there is a permutation $\sigma \in \mathfrak{S}_r$ such that: $\mathrm{ord}(c_1) = m_{\sigma(1)}, \ldots, \mathrm{ord}(c_r) = m_{\sigma(r)}$. We obtain that the datum of a surface isogenous to a higher product of unmixed type $S = (C_1 \times C_2)/G$ is determined, looking at the monodromy of each covering of C_i/G, by the datum of a finite group G together with an unmixed ramification structure. The condition (7) ensures that the action of G on the product of the two curves $C_1 \times C_2$ is free. We remark here that this can be specialized to $C_i/G \cong \mathbb{P}^1$ in order to obtain regular surfaces isogenous to a product. In this case condition (8) is automaticaly satisfied, see [16, Lemma 2.4]. Moreover, we can also ask $r_i = 3$, and therefore we obtain Beauville surfaces, in this case the ramification structure of G is called a *Beauville ramification structure*.

Remark 3.4 Note that a group G and an unmixed ramification structure (or equivalently a Beauville structure) determine the main invariants of the surface S. Indeed, by (5) and (4) we obtain:

$$4\chi(S) = |G| \cdot \left(2g_1' - 2 + \sum_{k=1}^{r_1} \left(1 - \frac{1}{m_{1,k}} \right) \right) \cdot \left(2g_2' - 2 + \sum_{k=1}^{r_2} \left(1 - \frac{1}{m_{2,k}} \right) \right), \tag{9}$$

and so, in the Beauville case,

$$4\chi(S) = 4(1 + p_g) = |G|(1 - \mu_1)(1 - \mu_2),$$

where

$$\mu_i := \frac{1}{m_{1,i}} + \frac{1}{m_{2,i}} + \frac{1}{m_{3,i}}, \quad (i = 1, 2). \tag{10}$$

The most important property of surfaces isogenous to a product is their weak rigidity property.

Theorem 3.5 ([10, Theorem 3.3, Weak Rigidity Theorem]) *Let $S = (C_1 \times C_2)/G$ be a surface isogenous to a higher product of curves. Then every surface with the same*

- *topological Euler number and*
- *fundamental group*

is diffeomorphic to S. The corresponding moduli space $\mathcal{M}^{top}(S) = \mathcal{M}^{diff}(S)$ of surfaces (orientedly) homeomorphic (resp. diffeomorphic) to S is either irreducible and connected or consists of two irreducible connected components exchanged by complex conjugation.

Thanks to the Weak Rigidity Theorem, we have that the moduli space of surfaces isogenous to a product of curves with fixed invariants—a finite group G and a type (τ_1, τ_2) in the unmixed case—consists of a finite number of irreducible connected components of $\mathcal{M}$. More precisely, let S be a surface isogenous to a product of curves of unmixed type with group G and a pair of disjoint systems of generators of type (τ_1, τ_2). By (9) we have $\chi(S) = \chi(G, (\tau_1, \tau_2))$, and consequently, by (5) $K_S^2 = K^2(G, (\tau_1, \tau_2)) = 8\chi(S)$, and $e(S) = e(G, (\tau_1, \tau_2)) = 4\chi(S)$. Moreover the fundamental group of S fits in the following exact sequence (cf. [8]):

$$1 \longrightarrow \pi_1(C_1) \times \pi_2(C_2) \longrightarrow \pi_1(S) \longrightarrow G \longrightarrow 1.$$

Let us fix a group G and a type (τ_1, τ_2) of an unmixed ramification structure, and denote by $\mathcal{M}_{(G,(\tau_1,\tau_2))}$ the moduli space of isomorphism classes of surfaces isogenous to a product of curves of unmixed type admitting these data, then it is obviously a subset of the moduli space $\mathcal{M}_{K^2(G,(\tau_1,\tau_2)),\chi(G,(\tau_1,\tau_2))}$. By the Weak Rigidity Theorem, the space $\mathcal{M}_{(G,(\tau_1,\tau_2))}$ consists of a finite number of irreducible connected components.

A group theoretical method to count the number of these components is given in [1, Theorem 1.3] in case of surfaces isogenous to a product of curves of unmixed type with $q = 0$ and G abelian. The following theorem is a natural generalization.

Theorem 3.6 ([22, Theorem 5.7]) *Let S be a surface isogenous to a product of unmixed type. Then we attach to S its finite group G (up to isomorphism) and the equivalence class ramification structures $(\mathcal{V}_1, \mathcal{V}_2)$ of type (τ_1, τ_2) of G, under the equivalence relation generated by:*

(1) Hurwitz moves and $\mathrm{Inn}(G)$ *on* $\mathcal{V}_1$,
(2) Hurwitz moves and $\mathrm{Inn}(G)$ *on* $\mathcal{V}_2$,
(3) simultaneous conjugation of $\mathcal{V}_1$ *and* $\mathcal{V}_2$ *by an element* $\phi \in \mathrm{Aut}(G)$, *i.e., we let* $(\mathcal{V}_1, \mathcal{V}_2)$ *be equivalent to* $(\phi(\mathcal{V}_1), \phi(\mathcal{V}_2))$.

Then two surfaces S and S' are deformation equivalent if and only if the corresponding pairs of systems of generators are in the same equivalence class.

If we fix a finite group G and a pair of types (τ_1, τ_2) of an unmixed ramification structure for G, counting the number of connected components of $\mathcal{M}_{(G,(\tau_1,\tau_2))}$ is then equivalent to the group theoretical problem of counting the number of classes of pairs of systems of generators of G of type (τ_1, τ_2) under the equivalence relation defined in Theorem 3.6. This leads also to the following definition.

Definition 3.7 Denote by $h(G; \tau_1, \tau_2)$ the number of *Hurwitz components*, namely the number of orbits of $\mathcal{U}(G; \tau_1, \tau_2)$ under the action of the group prescribed in Theorem 3.6.

4 Connected Components of the Moduli Space of Surfaces of General Type

We can simplify a lot the discussion of the previous section if we consider only *regular surfaces*, which will be assumed in the whole section. In this case, the Hurwitz moves are given only by the $\mathbf{t}_{\sigma_i}$ of Proposition 2.14, which corresponds to the ones described in Proposition 2.16. By Theorem 2.2 they are given by the braid group of the sphere on r strands $\mathbf{B}_r$ acting on the generators of the orbifold fundamental group as in Fig. 8.

In addition, we recall

Lemma 4.1 ([25, Lemma 9.4]) *The inner automorphism group,* $\mathrm{Inn}(G)$, *leaves each braid orbit invariant.*

This lemma allows us to use the above Theorem 3.6 in the simplified version without the $\mathrm{Inn}(G)$ action on the system of generators $\mathcal{V}_i$. Since the two actions of $\mathbf{B}_r$ and $\mathrm{Aut}(G)$ commute, one gets a double action of $\mathbf{B}_r \times \mathrm{Aut}(G)$ on the set of r-systems of generators for G.

Therefore, for regular surfaces isogenous to a product we have that fixing a finite group G and a pair of types $(\tau_1, \tau_2) = (m_{1,1}, \ldots, m_{1,r_1}, m_{2,1}, \ldots, m_{2,r_2})$ of an unmixed ramification structure for G counting the number of connected components of $\mathcal{M}_{(G,(\tau_1,\tau_2))}$ is then equivalent to the group theoretical problem of counting the number of classes of pairs of systems of generators of G of type (τ_1, τ_2) under the equivalence relation given by the action of $\mathbf{B}_{r_1} \times \mathbf{B}_{r_2} \times \mathrm{Aut}(G)$. In this case the number of Hurwitz components $h(G; \tau_1, \tau_2)$ is given by the number of orbits of $\mathcal{U}(G; \tau_1, \tau_2)$ under the following actions:

if $\tau_1 \neq \tau_2$ the action of $(\mathbf{B}_{r_1} \times \mathbf{B}_{r_2}) \times \mathrm{Aut}(G)$, given by:

$$((\gamma_1, \gamma_2), \phi) \cdot (T_1, T_2) := \big(\phi(\gamma_1(T_1)), \phi(\gamma_2(T_2))\big),$$

where $\gamma_1 \in \mathbf{B}_{r_1}, \gamma_2 \in \mathbf{B}_{r_2}, \phi \in \mathrm{Aut}(G)$ and $(T_1, T_2) \in \mathcal{U}(G; \tau_1, \tau_2)$.
if $\tau_1 = \tau_2$ the action of $(\mathbf{B}_r \wr \mathbb{Z}/2\mathbb{Z}) \times \mathrm{Aut}(G)$, where $\mathbb{Z}/2\mathbb{Z}$ acts on (T_1, T_2) by exchanging the two factors.

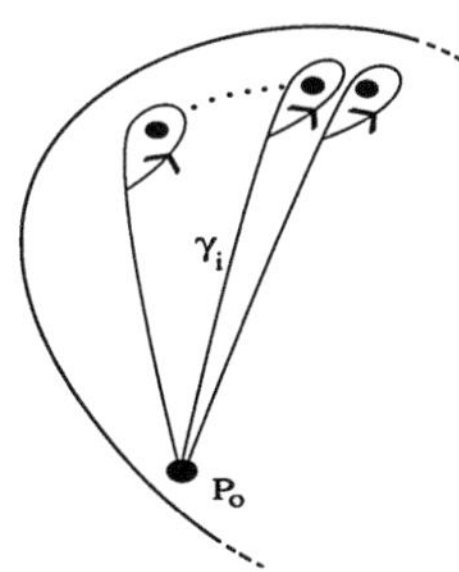

Fig. 8 Generators of of $\pi_1(\mathbb{P}^1 \setminus \mathcal{B}, p_0)$

In case of Beauville surfaces we define h as above substituting r_1 and r_2 with 3.

Proposition 4.2 *Fix r_1 and r_2 in $\mathbb{N}$. Let $\{G_n\}_{n=1}^{\infty}$ be a family of finite groups, which admit an unmixed ramification structure of size (r_1, r_2). Let $\tau_{n,1} = (m_{n,1,1}, \ldots, m_{n,1,r_1})$ and $\tau_{n,2} = (m_{n,2,1}, \ldots, m_{n,2,r_2})$ be sequences of types $(\tau_{n,1}, \tau_{n,2})$ of unmixed ramification structures for G_n, and $\{S_n\}_{n=1}^{\infty}$ be the family of surfaces isogenous to product with $q = 0$ admitting the given data, then as $|G_n| \overset{n \to \infty}{\longrightarrow} \infty$:*

(1) $\chi(S_n) = \Theta(|G_n|)$.
(2) $h(G_n; \tau_{n,1}, \tau_{n,2}) = O(\chi(S_n)^{r_1+r_2-2})$.

Proof (1) Note that, for $i = 1, 2$,

$$\frac{1}{42} \leq -2 + \sum_{j=1}^{r_i}\left(1 - \frac{1}{m_{n,i,j}}\right) \leq r_i - 2.$$

Indeed, for $r_i = 3$, the minimal value for $(1 - \mu_i)$ is $1/42$. For $r_i = 4$, the minimal value for $\left(-2 + \sum_{j=1}^{r_i}(1 - \frac{1}{m_{n,i,j}})\right)$ is $1/6$, and when $r_i \geq 5$, this value is at least $1/2$.
Now, by Eq. (9),

$$4\chi(S_n) = |G_n| \cdot \left(-2 + \sum_{j=1}^{r_1}\left(1 - \frac{1}{m_{n,1,j}}\right)\right) \cdot \left(-2 + \sum_{j=1}^{r_2}\left(1 - \frac{1}{m_{n,2,j}}\right)\right),$$

hence

$$\frac{|G_n|}{4 \cdot 42^2} \leq \chi(S_n) \leq \frac{(r_1 - 2)(r_2 - 2)|G_n|}{4}.$$

(2) For $i = 1, 2$, any r_i-system of generators $\mathcal{V}_{n,i}$ contains at most $r_i - 1$ independent elements of G_n. Thus, the size of the set of all unordered pairs of type $(\tau_{n,1}, \tau_{n,2})$ is bounded from above, by

$$|\mathcal{U}(G_n; \tau_{n,1}, \tau_{n,2})| \leq |G_n|^{r_1+r_2-2},$$

and so, the number of connected components is bounded from above by

$$h(G_n; \tau_{n,1}, \tau_{n,2}) \leq |G_n|^{r_1+r_2-2}.$$

Now, the result follows from (1).

Therefore we cannot expect more than a polynomial growth in the number of connected components of surfaces of general type if we count only regular surfaces isogenous to a product. Together with Shelly Garion we investigated the asymptotic growth of the number of connected components of the moduli space of surfaces of general type by using the group theoretical methods described above for regular surfaces isogenous to a product. Our first results already appeared on the ArXiv in 2009 (see arXiv:0910.5402v1). The following are some of the results contained in [16].

Notation 4.3 *Denote:*

- $h(n) = \Omega(g(n))$, *if* $h(n) \geq cg(n)$ *for some positive constant* c, *as* $n \to \infty$.
- $h(n) = \Theta(g(n))$, *if* $c_1 g(n) \leq h(n) \leq c_2 g(n)$ *for some positive constants* c_1, c_2, *as* $n \to \infty$.

Theorem 4.4 *Let* $\tau_1 = (m_{1,1}, \ldots, m_{1,r_1})$ *and* $\tau_2 = (m_{2,1}, \ldots, m_{2,r_2})$ *be two sequences of natural numbers such that* $m_{k,i} \geq 2$ *and* $\sum_{i=1}^{r_k}(1 - 1/m_{k,i}) > 2$ *for* $k = 1, 2$. *Let* $h(\mathfrak{A}_n, \tau_1, \tau_2)$ *be the number of connected components of the moduli space of surfaces isogenous to a product with* $q = 0$, *with group the alternating group* $\mathfrak{A}_n$, *and with type* (τ_1, τ_2). *Then*

$$(a) \quad h(\mathfrak{A}_n, \tau_1, \tau_2) = \Omega(n^{r_1 + r_2}),$$

and moreover,

$$(b) \quad h(\mathfrak{A}_n, \tau_1, \tau_2) = \Omega\big((\log(\chi))^{r_1 + r_2 - \epsilon}\big).$$

where $0 < \epsilon \in \mathbb{R}$.

Theorem 4.5 *Let* $\tau_1 = (m_{1,1}, \ldots, m_{1,r_1})$ *and* $\tau_2 = (m_{2,1}, \ldots, m_{2,r_2})$ *be two sequences of natural numbers such that* $m_{k,i} \geq 2$, *at least two of* $(m_{k,1}, \ldots, m_{k,r_k})$ *are even and* $\sum_{i=1}^{r_k}(1 - 1/m_{k,i}) > 2$, *for* $k = 1, 2$. *Let* $h(\mathfrak{S}_n, \tau_1, \tau_2)$ *be the number of connected components of the moduli space of surfaces isogenous to a product with* $q = 0$, *with group the symmetric group* $\mathfrak{S}_n$, *and with type* (τ_1, τ_2). *Then*

$$(a) \quad h(\mathfrak{S}_n, \tau_1, \tau_2) = \Omega(n^{r_1 + r_2}),$$

and moreover,

$$(b) \quad h(\mathfrak{S}_n, \tau_1, \tau_2) = \Omega\big((\log(\chi))^{r_1 + r_2 - \epsilon}\big).$$

where $0 < \epsilon \in \mathbb{R}$.

The proofs of part (a) of both Theorems are presented in [16, Sect. 3.2], and are based on results of Liebeck and Shalev [18]. The proofs of part (b) of both theorems appear in [16, Sect. 2].

We can specialize the results above to Beauville surfaces. Recall that a triple $(r, s, t) \in \mathbb{N}^3$ is said to be hyperbolic if

$$\frac{1}{r} + \frac{1}{s} + \frac{1}{t} < 1.$$

Corollary 4.6 *Let* $\tau_1 = (r_1, s_1, t_1)$ *and* $\tau_2 = (r_2, s_2, t_2)$ *be two hyperbolic types and let* $h(\mathfrak{A}_n, \tau_1, \tau_2)$ *be the number of Beauville surfaces with group* $\mathfrak{A}_n$ *and with types* (τ_1, τ_2). *Then*

$$(a) \quad h(\mathfrak{A}_n, \tau_1, \tau_2) = \Omega(n^6),$$

and moreover,

$$(b) \quad h(\mathfrak{A}_n, \tau_1, \tau_2) = \Omega\big((\log(\chi))^{6-\epsilon}\big).$$

where $0 < \epsilon \in \mathbb{R}$.

Corollary 4.7 *Let $\tau_1 = (r_1, s_1, t_1)$ and $\tau_2 = (r_2, s_2, t_2)$ be two hyperbolic types, assume that at least two of (r_1, s_1, t_1) are even and at least two of (r_2, s_2, t_2) are even, and let $h(\mathfrak{S}_n, \tau_1, \tau_2)$ be the number of Beauville surfaces with group $\mathfrak{S}_n$ and with types (τ_1, τ_2). Then*

$$(a) \quad h(\mathfrak{S}_n, \tau_1, \tau_2) = \Omega(n^6),$$

and moreover,

$$(b) \quad h(\mathfrak{S}_n, \tau_1, \tau_2) = \Omega\big((\log(\chi))^{6-\epsilon}\big).$$

where $0 < \epsilon \in \mathbb{R}$.

The situation is more interesting for abelian groups. We have the following results which assure the existence of ramification structure for abelian group.

Theorem 4.8 *Let G be an abelian group, given as*

$$G \cong \mathbb{Z}/n_1\mathbb{Z} \times \cdots \times \mathbb{Z}/n_t\mathbb{Z},$$

where $n_1 \mid \cdots \mid n_t$. For a prime p, denote by $l_i(p)$ the largest power of p which divides n_i (for $1 \leq i \leq t$).

Let $r_1, r_2 \geq 3$, then G admits an unmixed ramification structure of size (r_1, r_2) if and only if the following conditions hold:

- $r_1, r_2 \geq t + 1$;
- $n_t = n_{t-1}$;
- *If $l_{t-1}(3) > l_{t-2}(3)$ then $r_1, r_2 \geq 4$;*
- $l_{t-1}(2) = l_{t-2}(2)$;
- *If $l_{t-2}(2) > l_{t-3}(2)$ then $r_1, r_2 \geq 5$ and r_1, r_2 are not both odd.*

This theorem is proved in [16, Sect. 3.4]. Once we established the existence of ramification structures, we can put the surfaces in sequences and compute the number of connected components of the moduli space. More precisely, the following holds.

Theorem 4.9 *Let $\{S_p\}$ be the family of surfaces isogenous to a product with $q = 0$ with group $G_p := (\mathbb{Z}/p\mathbb{Z})^r$ admitting ramification structure of type $\tau_p = (p, \ldots, p)$ (p appears $(r + 1)$–times) where p is prime. If we denote by $h(G_p; \tau_p, \tau_p)$ the number of connected components of the moduli space of isomorphism classes of surfaces isogenous to a product with $q = 0$ admitting these data, then*

$$h(G_p; \tau_p, \tau_p) = \Theta(\chi^r(S_p)).$$

Therefore, there exist families of surfaces such that the degree of the polynomial h in χ (and so in K^2) can be arbitrarily large. The proof of this theorem appears in [16, Sect. 2]. Notice that not only the number of connected components increases, but also their sizes. Indeed, we see that $2r - 6$ is the dimension of these connected components.

Again we can specialize the results for Beauville surfaces.

Corollary 4.10 *Let* $\{S_p\}$ *be the family of Beauville surfaces* $G_p := (\mathbb{Z}/p\mathbb{Z})^2$ *admitting ramification structure of type* $\tau_p = (p, p, p)$ *where* $p \geq 5$ *is prime. If we denote by* $h(G_p; \tau_p, \tau_p)$ *the number of Beauville surfaces admitting these data. Then*

$$h(G_p; \tau_p, \tau_p) = \Theta(\chi^2(S_p)).$$

Proof Let $(x_1, x_2; y_1, y_2)$ be an unmixed Beauville structure for G. Since x_1, x_2 are generators of G, they are a basis, and without loss of generality x_1, x_2 are the standard basis $x_1 = (1, 0)$, $x_2 = (0, 1)$. Now, let $y_1 = (a, b)$, $y_2 = (c, d)$, then the condition (7) means that any pair of the six vectors yield a basis of G, implying that a, b, c, d must satisfy the following conditions

$$a - b, a + c, c - d, b + d, a + c - b - d, ad - bc \in U \tag{11}$$

Moreover, the number N_p of quadruples (a, b, c, d) satisfy (11) is $N_p = (p - 1)$ $(p - 2)(p - 3)(p - 4)$. The pairs $\big((1, 0), (0, 1); (a, b), (c, d)\big)$, where a, b, c, d satisfy (11), are exactly the representatives for the $\mathrm{Aut}(G)-$orbits in the set $\mathcal{U}(G; \tau, \tau)$.

Now, one should consider the action of $\mathbf{B}_3 \times \mathbf{B}_3$ on $\mathcal{U}(G; \tau, \tau)$, which is equivalent to the action of $\mathfrak{S}_3 \times \mathfrak{S}_3$, since G is abelian. The action of $\mathfrak{S}_3$ on the second component is obvious (there are 6 permutations), and the action of $\mathfrak{S}_3$ on the first component can be translated to an equivalent $\mathrm{Aut}(G)-$action, given by multiplication in one of the six matrices:

$$\begin{pmatrix} 1 & 0 \\ 0 & 1 \end{pmatrix}, \begin{pmatrix} 0 & 1 \\ 1 & 0 \end{pmatrix}, \begin{pmatrix} -1 & 0 \\ -1 & 1 \end{pmatrix}, \begin{pmatrix} 1 & -1 \\ 0 & -1 \end{pmatrix}, \begin{pmatrix} -1 & 1 \\ -1 & 0 \end{pmatrix}, \begin{pmatrix} 0 & -1 \\ 1 & -1 \end{pmatrix},$$

yielding an equivalent representative.

Therefore, the action of $\mathfrak{S}_3$ on the second component yields orbits of length 6, and the action of $\mathfrak{S}_3$ on the first component connects them together, and gives orbits of sizes from 6 to 36. Moreover since one can exchange the vector (x_1, x_2) with the vector (y_1, y_2) we get

$$N_p/72 \leq h \leq N_p/6.$$

By Proposition 4.2, we have as $p \to \infty$:

$$\chi(S_p) = \Theta(p^2),$$

while by the above computation we have

$$h(G_p; \tau_p, \tau_p) = \Theta(p^4).$$

Therefore

$$h(G_p; \tau_p, \tau_p) = \Theta(\chi^2(S_p)).$$

Corollary 4.11 *Let n be an integer such that $(n, 6) = 1$. The number $h = h(G; \tau, \tau)$, where $\tau = (n, n, n)$, of Hurwitz components for $G = (\mathbb{Z}/n\mathbb{Z})^2$, where $n = p_1^{k_1} \cdot \ldots \cdot p_t^{k_t}$, satisfies*

$$N_n/72 \le h \le N_n/6,$$

where $N_n = \prod_{i=1}^{t} p_i^{4k_i - 4}(p_i - 1)(p_i - 2)(p_i - 3)(p_i - 4)$.

Remark 4.12 Notice that if n is divisible by the first l primes $p_i \ge 5$ then since:

$$\lim_{l \to \infty} \prod_i (1 - \frac{1}{p_i}) = 0$$

we have $N_n/n^4 \to 0$ as $l \to \infty$.

In [17] the authors give an explicit formula for the number of isomorphism classes of Beauville surfaces $\Theta(n)$, which we now explain. We shall keep the notation of [17]. Define the following functions for $n, e \in \mathbb{N}$ and p prime.

$$\Theta_1(n) := n^4 \prod_{p \mid n} \left(1 - \frac{1}{p}\right) \left(1 - \frac{2}{p}\right) \left(1 - \frac{3}{p}\right) \left(1 - \frac{4}{p}\right);$$

$$\Theta_2(p^e) := \begin{cases} p^{2e} \left(1 - \frac{1}{p}\right) \left(1 - \frac{2}{p}\right) & if \quad p \equiv 1 \mod 4, \\ p^{2e} \left(1 - \frac{1}{p}\right) \left(1 - \frac{4}{p}\right) & if \quad p \equiv 3 \mod 4; \end{cases}$$

$$\Theta_3(p^e) := p^{2e} \left(1 - \frac{3}{p}\right) \left(1 - \frac{5}{p}\right);$$

$$\Theta_4(p^e) := \begin{cases} 0 & if \quad p \equiv -1 \mod 3, \\ 2 & if \quad p \equiv 1 \mod 3. \end{cases}$$

Theorem 4.13 ([17, Theorem 2]) *Let $n = p_1^{e_1} \cdot \ldots \cdot p_t^{e_t}$ be an integer such that $(n, 6) = 1$. Then the number of isomorphism classes of Beauville surfaces with group $G = (\mathbb{Z}/n\mathbb{Z})^2$ is*

$$\Theta(n) := \frac{1}{72} \left(\Theta_1(n) + 4 \prod_{i=1}^{t} \Theta_2(p_i^{e_i}) + 6 \prod_{i=1}^{t} \Theta_3(p_i^{e_i}) + 12 \prod_{i=1}^{t} \Theta_4(p_i^{e_i}) \right).$$

The case of *irregular* surfaces has not been so intensively investigated. Indeed, counting the number of connected components of the moduli space of surfaces isogenous to a product with fixed data can be more complicated. We have to consider Theorem 3.6 together with the Hurwitz moves described in Proposition 2.14. Indeed, this procedure was applied in two specific cases: For surfaces isogenous to a product with $p_g = q = 1$ or 2 in [22, 24]. In both cases it was used a GAP4 script which can be found in [23]. It would be interesting to consider all the surfaces isogenous to a product with fixed invariants K^2 and χ and count the number of connected components of the moduli space of surfaces of general type they give. Indeed for $K^2 = 8$ and $\chi = 1$ we can count already 94 connected componens. Respectively: 1 component with $p_g = q = 4$, 1 with $p_g = q = 3$, 27 with $p_g = q = 2$, 52 with $p_g = q = 1$ and 13 with $p_g = q = 0$.

In this paper we consider only *unmixed* surfaces isogenous to a product, nevertheless similar methods could be applied also for the *mixed* case. Indeed, there are already works in these direction, see [14].

Acknowledgments The author is grateful to G. Bini for reading and commenting the paper. Moreover the author thanks the organizers of the conference *Beauville surfaces and Groups* N. Barker, I. Bauer, S. Garion and A. Vdovina for the invitation and the kind hospitality.

References

1. I. Bauer, F. Catanese, Some new surfaces with $p_g = q = 0$, in *Proceeding of the Fano Conference*, (Torino, 2002), pp. 123–142
2. I. Bauer, F. Catanese, F. Grunewald, Beauville surfaces without real structures, in *Geometric Methods in Algebra and Number Theory*, Progress in Mathematics, vol. 235 (Birkhäuser, Boston, 2005), pp. 1–42
3. I. Bauer, F. Catanese, F. Grunewald, Chebycheff and Belyi polynomials, dessins d'enfants, Beauville surfaces and group theory. Mediterr. J. Math. **3**, 121–146 (2006)
4. I. Bauer, F. Catanese, F. Grunewald, The classification of surfaces with $p_g = q = 0$ isogenous to a product. Pure Appl. Math. Q. **4**, 547–586 (2008)
5. J.S. Birman, Mapping class groups and their relationship to braid groups. Commun. Pure Appl. Math. **XXII**, 213–238 (1969)
6. J.S. Birman, *Braids, Links, and Mapping Class Groups*, vol. 82, Annals of Mathematics Studies (Princeton University Press, Princeton, 1974)
7. F. Catanese, Chow varieties, Hilbert schemes and moduli spaces of surfaces of general type. J. Algebraics Geom. **1**, 561–595 (1992)
8. F. Catanese, Fibred surfaces, varieties isogenous to a product and related moduli spaces. Am. J. Math. **122**, 1–44 (2000)
9. F. Catanese, Differentiable and deformation type of algebraic surfaces, real and symplectic structures. Symplectic 4-Manifolds and Algebraic Surfaces. Lectures Given at the C.I.M.E. Summer School, Cetraro, Italy, 2–10 September 2003, Lecture Notes in Mathematics, vol. 1938 (Springer, Berlin, 2008), pp. 55–167
10. F. Catanese, Moduli spaces of surfaces and real structures. Ann. Math. **158**, 577–592 (2003)
11. F. Catanese, Trecce, mapping class groups, fibrazioni di Lefschetz e applicazioni al diffeomorfismo di superfici algebriche. Luigi Cremona (1830–1903) (Incontri Studio 36., Milano, 2005), pp. 207–235

12. F. Catanese, M. Lönne, F. Perroni, Irreducibility of the space of dihedral cover of the projective line of a given numerical type. arXiv:1206.5498
13. M. Dehn, Die Gruppe der Abbildungsklassen. (Das arithmetische Feld auf Flächen.). Acta Math. **69**, 135–206 (1938)
14. D. Frapporti, R. Pignatelli, Mixed quasiale quotients with arbitrary singularities. Glasg. Math. J. **57** 143–165 (2015)
15. S. Garion, M. Penegini, New Beauville surfaces and finite simple groups. Manuscripta Math. **142** 391–408 (2013)
16. S. Garion, M. Penegini, Beauville surfaces, moduli spaces and finite groups. Comm. Algebra. **42**, 2126–2155 (2014)
17. G. Gonzales-Diez, G. Jones, D. Torres-Teigell, Beauville surfaces with abelian Beauville group. Preprint. arXiv:1102.4552v3
18. M.W. Liebeck, A. Shalev, Fuchsian groups, coverings of Riemann surfaces, subgroup growth, random quotients and random walks. J. Algebra **276**, 552–601 (2004)
19. M.W. Liebeck, A. Shalev, Fuchsian groups, finite simple groups and representation varieties. Invent. Math. **159**(2), 317–367 (2005)
20. C. Maclachlan, Modular groups and fiber spaces over Teichmüller spaces, in *Proceedings of the 1973 Conference on Discontinuous Groups Riemann Surf.*, University of Maryland, pp. 297–314 (1974)
21. M. Manetti, Iterated double covers and connected components of moduli spaces. Topology **36**, 745–764 (1997)
22. M. Penegini, The classification of isotrivially fibred surfaces with $p_g = q = 2$. With an Appendix of S. Rollenske. Collect. Math. **62**, 239–274 (2011)
23. M. Penegini, The classification of isotrivially fibred surfaces with $p_g = q = 2$, and topics on Beauville surfaces. Ph.D. Thesis, Universität Bayretuh (2010)
24. F. Polizzi, On surfaces of general type with $p_g = q = 1$ isogenous to a product of curves. Commun. Algebra **36**, 2023–2053 (2008)
25. H. Völklein, *Groups as Galois Groups—An Introduction*, vol. 53, Cambridge Studies in Advanced Mathematics (Cambridge University Press, Cambridge, 1996)

On Quasi-Étale Quotients of a Product of Two Curves

Roberto Pignatelli

Abstract A quasi-étale quotient of a product of two curves is the quotient of a product of two curves by the action of a finite group which acts freely out of a finite set of points. A quasi-étale surface is the minimal resolution of the singularities of a quasi-étale quotient. They have been successfully used in the last years by several authors to produce several interesting new examples of surfaces. In this paper we describe the principal results on this class of surfaces, and report the full list of the minimal quasi-étale surfaces of general type with geometric genus equal to the irregularity ≤ 2.

1 Introduction

Throughout this paper a "surface" (resp. curve) is a smooth complex algebraic surface (resp. curve); these are compact complex manifolds of dimension 2 (resp. 1) with an algebraic structure.

We are interested in the birational geometry of surfaces; in other words we look at surfaces modulo the equivalence relation generated by the blow-up in a point.

For sake of simplicity, we will restrict to projective surfaces, so we assume that the surface can be algebraically embedded in a projective space. Most of the statements in this section are classical results in complex algebraic geometry; a good reference is the classical book [3].

For each surface S, we will denote

- by Ω_S^q the sheaf of algebraic q-forms
- by $\mathcal{O}_S$ the structure sheaf Ω_S^0

The author was partially supported by the projects PRIN2010-2011 Geometria delle varietà algebriche and Futuro in Ricerca 2012 Moduli Spaces and Applications.

R. Pignatelli (✉)
Dipartimento di Matematica dell'Università di Trento,
Via Sommarive 14, 38123 Trento, TN, Italy
e-mail: Roberto.Pignatelli@unitn.it

© Springer International Publishing Switzerland 2015

I. Bauer et al. (eds.), *Beauville Surfaces and Groups*, Springer Proceedings in Mathematics & Statistics 123, DOI 10.1007/978-3-319-13862-6_10

- by K_S a canonical divisor, that is the divisor of the zeroes and of the poles of a meromorphic algebraic 2-form
- for each divisor D in S, by $\mathcal{O}_S(D)$ the invertible sheaf associated to it; so $\mathcal{O}_S(K_S) \cong \Omega_S^2$
- for each sheaf $\mathcal{F}$, by $h^q(\mathcal{F})$ the dimension of the qth Cech cohomology group $H^q(\mathcal{F})$.

The "classical" birational invariants are

- the topological fundamental group $\pi_1(S)$
- the geometric genus

$$p_g(S) := h^{2,0}(S) := h^0(\Omega_S^2) = h^0(\mathcal{O}_S(K_S)) = h^2(\mathcal{O}_S)$$

- the irregularity

$$q(S) := h^{1,0}(S) := h^0(\Omega_S^1) = h^1(\mathcal{O}_S)$$

- the Euler characteristic

$$\chi(\mathcal{O}_S) := 1 - q + p_g$$

- the plurigenera

$$P_n(S) := h^0(\mathcal{O}_S(nK_S))$$

- the Kodaira dimension $\kappa(S)$, which is the smallest number κ such that $\frac{P_n}{n^\kappa}$ is bounded from above, known to be at most 2.

Definition 1.1 A surface S is of *general type* if $\kappa(S) = 2$.

The Enriques-Kodaira classification provides a relatively good understanding of the surfaces of *special* type, which are those with Kodaira dimension $\kappa(S) \neq 2$ (this includes rational surfaces, K3 surfaces, Enriques surfaces, Abelian surfaces, elliptic surfaces...). The class of the surfaces of general type may be considered as the class of the surfaces... we do not understand, and therefore the most interesting, at least in the opinion of the author. It is worth mentioning that the method we are going to discuss has been very recently applied also to a different (very interesting) class of surfaces, the K3 surfaces, see [17].

Definition 1.2 A surface S is minimal if K_S is nef, that is if the intersection of K_S with every curve is nonnegative.

In the birational class of a surface of general type S there is exactly one minimal surface, say $\bar{S}$, which is *its minimal model*. The self intersection of the canonical divisor $K_{\bar{S}}^2$, which is not a birational invariant, measures in some sense the distance among S and $\bar{S}$; more precisely S is obtained by $\bar{S}$ by a sequence of exactly $K_{\bar{S}}^2 - K_S^2$ blow ups.

If S is of general type, then the Riemann-Roch formula computes all plurigenera $P_n(S)$ from $\chi(\mathcal{O}_S)$ and $K_{\bar{S}}^2$; so knowing p_g, q, $K_{\bar{S}}^2$ and the topological fundamental group is enough to have all the birational invariants mentioned above.

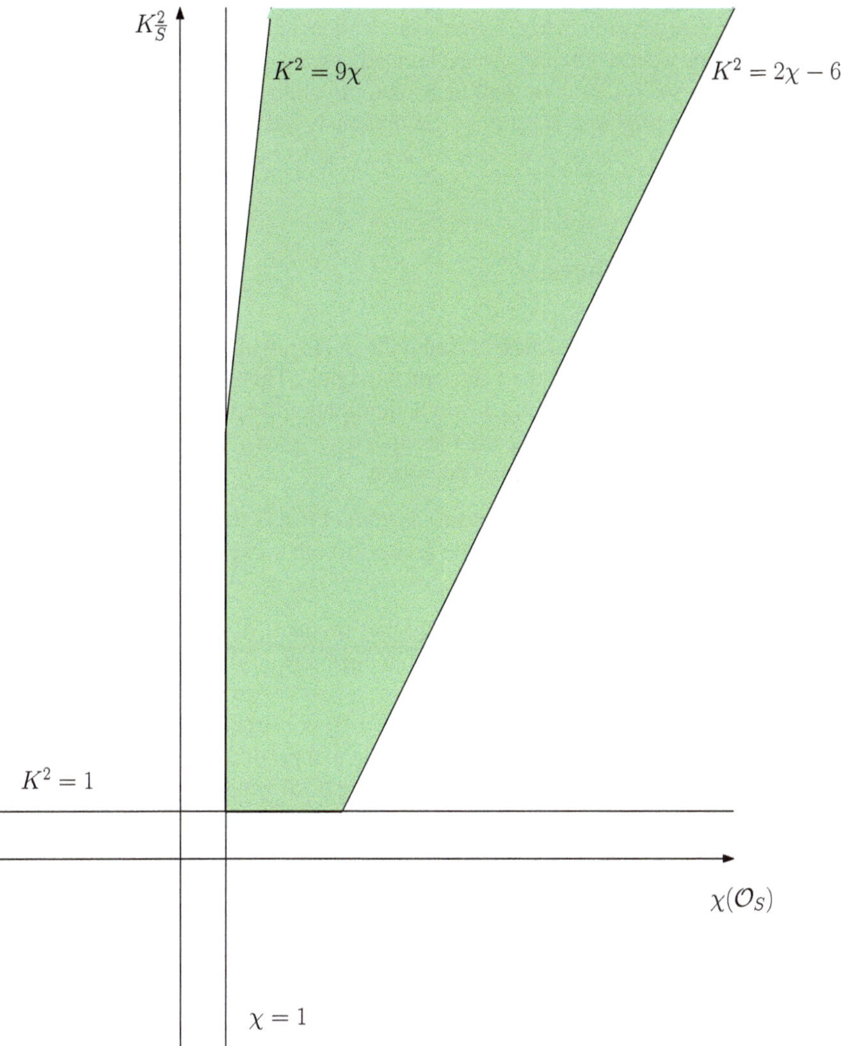

Fig. 1 The geography of the surfaces of general type

This leads us to a famous picture, known as *the geography* of the surfaces of general type. By some famous inequalities (Noether, Bogomolov-Miyaoka-Yau,...) the possible values of the pair $(\chi,\ K_S^2)$ are in the green region of the Fig. 1.

It is still unknown if all integral points in the green region of the Fig. 1 can be obtained by surfaces of general type, although we know we can fill "most" of it: this is an example of *geographical question*. More generally, one would like to know all possible values of the 4-tuple of topological invariants $(p_g(S), q(S), K_S^2, \pi_1(S))$; we are very far from that, but one can hope to answer this question at least for some portion of the green region in Fig. 1. Some of the most interesting surfaces of general type, for reasons we are not going to explain here (see e.g. [6, 8]), lie on the vertical "boundary" line $\chi(\mathcal{O}_S) = 1$; we will be back to this line at the end of the paper.

To answer geographical questions it is obviously very important to have some tools to construct examples, as the one we are going to explain.

2 Quasi-Étale Surfaces

The idea starts from a construction of Beauville in [11], which gives the name to the Beauville surfaces which are the center of this book. This is a minimal surface of general type with $p_g = q = 0$ and $K^2 = 8$, quotient of the product of two curves of genus 6 by a free action of the Abelian group $\mathbb{Z}^2_{/5\mathbb{Z}}$.

This example led to the following definition

Definition 2.1 A surface is *isogenous to a product* if it is the quotient of a product of two curves by a free action of a finite group. If both curves have genus at least 2, then it is isogenous to a *higher* product.

Beauville surfaces are isogenous to a higher product. A surface of general type isogenous to a product is automatically isogenous to a higher product, so we will drop the word 'higher' in the following.

Several authors (e.g. [4, 5, 13, 18, 20, 23, 24]) constructed then new examples of surfaces of general type as surfaces isogenous to a product, in particular surfaces with $p_g = q = 0$ as Beauville example. All surfaces of general type isogenous to a product are minimal with $K^2 = 8\chi$, which forces them in a line of the 2-dimensional Fig. 1. This is a strong limitation from the point of view of the birational geometry of surfaces of general type, since it shows that the construction of surfaces isogenous to a product can answer only very particular geographical questions.

This suggested to weaken Definition 2.1, to get something which is as simple to construct, but not limited by $K^2 = 8\chi$. A possibility is the following:

Definition 2.2 A *quasi-étale quotient* is the quotient of a product of two curves by the action of a finite group G acting freely out of a finite set of points. A *quasi-étale surface* is the minimal resolution of the singularities of a quasi-étale quotient.

Indeed, a quasi-étale quotient is smooth if and only if the action is free, so the surface is isogenous to a product. Each point in the product of the two curves stabilized by a non trivial subgroup of G maps onto a singular point of the quotient. We will always denote by X the singular quotient, and by S the smooth resolution of its singularities.

We have in this case the additional problem to study the singularities of X and their minimal resolution. The advantage is that these surfaces get out of the line $K^2 = 8\chi$. Indeed, it is easy to prove that $K^2 \leq 8\chi$, but apparently no other constraint applies [26] proves $K^2_S \neq 8\chi - 1$ for a quasi-étale unmixed surface, but this does not extend to the mixed case) and we may hope to fill the whole yellow region in Fig. 2.

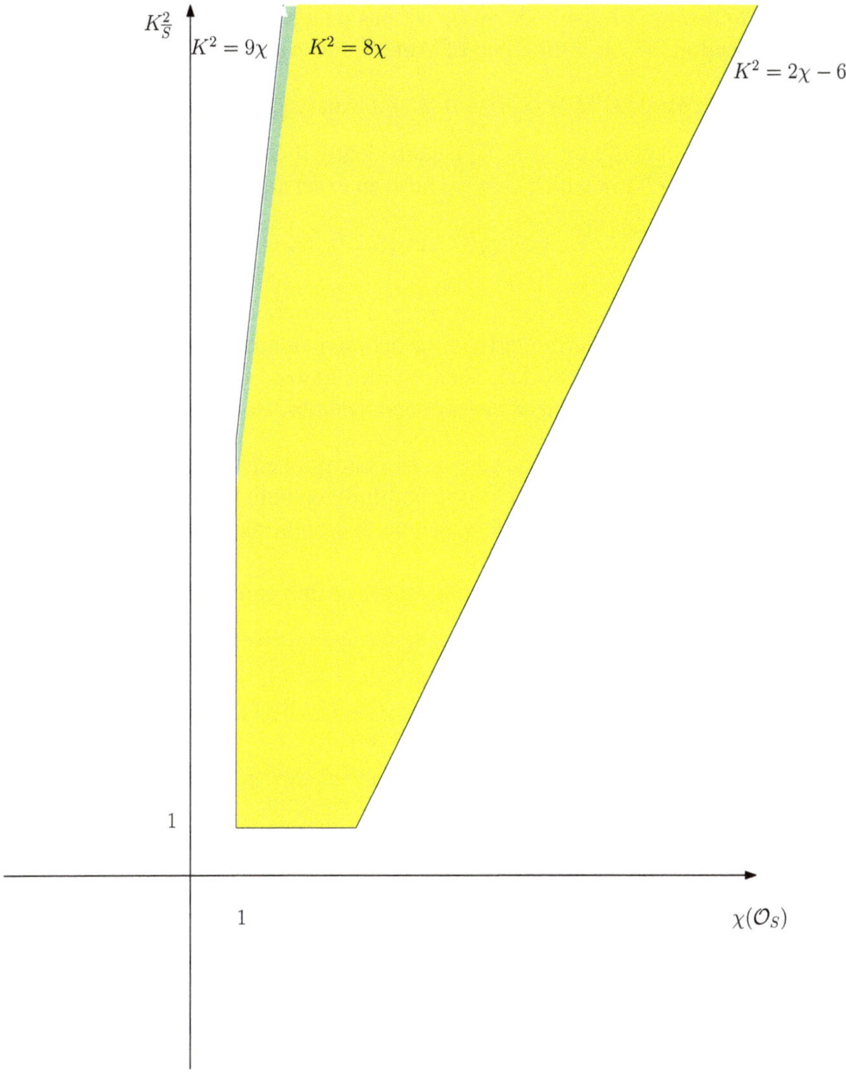

Fig. 2 The region of the geography we expect to be able to fill with quasi-étale surfaces

Quasi-étale surfaces splits naturally in two classes. Indeed (see [14]), if C_1 and C_2 are two algebraic curves

- either C_1 and C_2 are not isomorphic, in which case $\mathrm{Aut}\,(C_1 \times C_2) \cong \mathrm{Aut}\,(C_1) \times \mathrm{Aut}\,(C_2)$
- or $C_1 \cong C_2 \cong C$: in this case $\mathrm{Aut}\,(C \times C) = (\mathrm{Aut}\,C)^2 \rtimes \mathbb{Z}_2$, where the involution generating the (non normal) subgroup on the right is the exchange of the factors.

This leads to distinguish the finite group actions on a product of two curves in two classes, depending if G is a subgroup of $\mathrm{Aut}(C_1) \times \mathrm{Aut}(C_2)$ or not.

Definition 2.3 We set $G^{(0)} = G \cap (\mathrm{Aut}(C_1) \times \mathrm{Aut}(C_2))$, and say that the action is

unmixed if $G < \mathrm{Aut}(C_1) \times \mathrm{Aut}(C_2)$, equivalently if $G = G^{(0)}$.
  mixed if $G \neq G^{(0)}$, in which case we have an exact sequence

$$1 \to G^{(0)} \to G \to \mathbb{Z}_2 \to 1 \tag{1}$$

We will say that a quasi-étale quotient, resp. a quasi-étale surface is mixed if the action defining it is mixed. Similarly, if the action is unmixed, we will say that the induced quasi-étale quotient/surface is unmixed. Unmixed quasi-étale surfaces have been also called *product-quotient surfaces* and *standard isotrivial fibrations*.

By a remark of Catanese [14], every quasi-étale quotient can be constructed by a minimal action, that means that $G^{(0)}$ acts faithfully on both factors. So without loss of generality we can and will always assume that both maps $G^{(0)} \to \mathrm{Aut}\, C_i$ are injective.

In this language, the quasi-étale condition has a completely algebraic surprisingly simple description.

Theorem 2.4 ([15]) *Consider a minimal action of a finite group G on a product of two curves. Then G do not act freely out of a finite set of points if and only if the action is unmixed and the exact-sequence (1) splits.*

In other words G acts freely out of a finite set of points if and only if either the action is unmixed or the exact sequence (1) does not split.

In other words, the only minimal actions which violate the quasi-étale condition are the mixed actions in which $G \backslash G^{(0)}$ contains an involution.

Remark 2.5 Indeed, if $G \backslash G^{(0)}$ contains an involution, then one can assume (up to automorphisms) that the involution is $(x, y) \mapsto (y, x)$, which fixes the diagonal, and therefore the action is not free out of a finite set of points. Theorem 2.4 says that this is the only case in which the quasi-étale condition fail. In this last case, $(C \times C)/G$ is dominated by the symmetric product $C^{(2)}$ of the curve C. It would be interesting to extend to this case all results we have for the quasi-étale case.

3 Constructing Curves with a Finite Group Action

In order to construct a quasi-étale surface, we need curves with several automorphisms; indeed the general curve of genus $g > 2$ has no nontrivial automorphisms at all. It is worth mentioning here that if $g \geq 2$, which is the case we are interested in, $\mathrm{Aut}\, C$ is finite by the classical Schwarz Theorem. Anyway the method we are going to describe works also for $g \leq 1$, assuming the order of the group to be finite.

The method for constructing it, based on the classical Riemann Existence Theorem, is the following.

Let C be a smooth algebraic curve, and let $G^{(0)}$ be a finite subgroup of $\mathrm{Aut}\,C$. Then $C' := C/G$ is a smooth algebraic curve of smaller genus g' and the projection $C \to C'$ has finitely many critical values, say $p_1, \ldots, p_r$. Let us fix one of these p_i; each of its preimages is stabilized by a cyclic subgroup of $G^{(0)}$, and all these subgroups lie in the same conjugacy class; in particular they all have the same cardinality, say m_i.

Removing the p_i from C' and their preimages from C, we remain with a regular topological cover $C^0 \to C'\backslash\{p_i\}$; these are determined by their monodromy map $\pi_1(C'\backslash\{p_i\}) \to G^{(0)}$. The Riemann Existence Theorem shows that each such topological cover may be uniquely extended to a map among compact complex curves $C \to C'$.

This gives a way to construct all pairs $(C, G^{(0)})$ where C is a smooth compact complex curve and $G^{(0)}$ is a finite subgroup of $\mathrm{Aut}\,C$.

Theorem 3.1 (Consequence of the Riemann Existence Theorem) *Given*

(a) *a compact complex curve C'*
(b) *a finite set $\{p_i\} \subset C'$*
(c) *a surjective homomorphism $\pi_1(C'\backslash\{p_i\}) \to G^{(0)}$*

there is, up to automorphisms, a unique curve C, and a unique inclusion of $G^{(0)}$ in $\mathrm{Aut}\,C$ such that $C' = C/G^{(0)}$, the critical values of the quotient map $C \to C'$ belong to the set $\{p_i\}$, and such that, removing $\{p_i\}$ from C' and its preimage from C, we get the topological cover whose monodromy map is the map in (c).

The key point here is the homomorphism (c). The more effective way is to construct a *generating vector* (see, e.g., [26]) of $G^{(0)}$. One chooses in a "standard" way a set of generators of $\pi_1(C'\backslash\{p_i\})$, and then gives the map (c) by giving their images in $G^{(0)}$; we need then a set of generators (to ensure the surjectivity) of $G^{(0)}$ respecting some relations reflecting the relations among the chosen generators of $\pi_1(C'\backslash\{p_i\})$.

These relations are summarized in the definition of *generating vector* of *signature* $(g'; m_1, \ldots, m_r)$. We do not repeat here this definition, referring the interested reader, e.g., to [16]; just repeat here that g' is the genus of C', r the number of critical values of the map $C \to C'$, m_i the order of the stabilizer of each preimage of the critical value p_i.

By Hurwitz formula, the genus of C can be computed by the signature of the generating formula, namely

$$2g(C) - 2 = |G|\left(2g' - 2 + \sum_i \left(1 - \frac{1}{m_i}\right)\right) \qquad (2)$$

4 Constructing Quasi-Étale Quotients

In the case of unmixed actions (where, as seen in Theorem 2.4, the quasi-étale condition is empty), we need to give two curves C_1, C_2 and inclusions of $G^{(0)}$ in both $\mathrm{Aut}\,(C_i)$. Following the strategy in Sect. 3 we need to give two generating vectors of $G^{(0)}$, look at their signatures, say $(g'_1; m_1, \ldots, m_r)$ and $(g'_2; n_1, \ldots, n_s)$, and then choose two curves C'_1, C'_2 of respective genera g'_1, g'_2, and two finite subsets $\{p_1, \ldots, p_r\} \subset C'_1$, $\{q_1, \ldots, q_s\} \subset C'_2$.

In the mixed case, the construction is even simpler; one needs only one generating vector of $G^{(0)}$, look at its signature, say $(g'; m_1, \ldots, m_r)$, then choose a curve C' of genus g' and a finite subset $\{p_1, \ldots, p_r\} \subset C'$, and finally choose a degree 2 extension G of $G^{(0)}$ as in (1).

These data determine (see [14, 15]) a mixed action on the product of two curves as follows. First of all they give, as in Sect. 3, a curve C and an inclusion of $G^{(0)}$ in $\mathrm{Aut}\,C$. We choose an element $\tau' \in G \backslash G^{(0)}$, and notice that for each element $g \in G \backslash G^{(0)}$ there is a unique $g_0 \in G^{(0)}$ such that $g = \tau'g_0$. Then we give the mixed action on $C \times C$ as follows: we note that $(\tau')^2 \in G^{(0)}$ and set, $\forall g_0 \in G^{(0)}, \forall x, y \in C$

$$\begin{cases} g_0(x, y) = & (g_0 x, \tau' g_0 \tau'^{-1} y) \\ \tau' g_0(x, y) = & (\tau' g_0 \tau'^{-1} y, \tau'^2 g_0 x) \end{cases}$$

In the mixed case, since we are only interested in the quasi-étale surfaces, we will assume from now on that the extension (1) is unsplit.

4.1 Singularities

We have a recipe to construct quasi-étale quotient; to obtain the quasi-étale surfaces, we need to understand their singularities.

Let then $X := (C_1 \times C_2)/G$ be a quasi-étale quotient. Then [9, 15, 16, 19] the singularities of X are the images of all points in $C_1 \times C_2$ which are stabilized by some nontrivial subgroup of G; these can be computed by the generating vectors, see, e.g., [9] for the unmixed case and [16] for the mixed case. We recall here the analytic type of singularities one can find.

Definition 4.1 For each rational number $0 < \frac{q}{n} < 1$ $(\gcd(q, n) = 1)$ a singularity of type $C_{n,q}$ (also called of type $\frac{1}{n}(1, q)$, or of type $\frac{q}{n}$) is an isolated singularity locally isomorphic to the singularity obtained by quotienting $\mathbb{C}^2$ by the action of the diagonal matrix with eigenvalues $e^{\frac{2\pi i}{n}}$ and $e^{\frac{2q\pi i}{n}}$.

The exceptional divisor of the minimal resolution of a singularity of type $C_{n,q}$ is a chain of rational curves $A_1, \ldots, A_k$, each intersecting only the previous and the next one transversally in a single point, with respective self intersections $-b_1, \ldots, -b_k$ where the b_i are given by the continued fraction

$$\frac{n}{q} = [b_1, \ldots, b_k] = b_1 - \cfrac{1}{b_2 - \cfrac{1}{b_3 - \cdots}}.$$

In other words, the dual graph of the exceptional divisor is

$-b_1 \qquad -b_2 \qquad\qquad\qquad\qquad\qquad\qquad\qquad -b_{k-1} \quad -b_k$

•——————•- -•——————•

In the mixed case, we have an intermediate unmixed quotient $Y := C^2/G^{(0)}$ which has then only cyclic quotient singularities, and a double cover $Y \to X$, so we can see X as the quotient of Y by an involution.

Since the square of each element of $G \backslash G^{(0)}$ is a nontrivial element of $G^{(0)}$, the points of $C \times C$ stabilized by a nontrivial subgroup of $G^{(0)}$ are the same stabilized by a nontrivial subgroup of G; therefore the singular locus of X is the image of the singular locus of Y and, once we have computed the cyclic quotient singularities of Y, we only have to describe how the involution acts on them.

There are two possibilities. If the involution exchange two singular points, then they are isomorphic, and we get one singularity of the same type $C_{n,q}$ on the quotient. In the other case the following holds

Theorem 4.2 ([16]) *Let P be a singular point of Y which is fixed by the involution. Then P is a singular point of type $C_{n,q}$ with $q^2 \equiv 1 \bmod n$; in other words the dual graph of the minimal resolution of the singularity of P is a symmetric string*

$-b_1 \qquad -b_2 \qquad\qquad\qquad\qquad\qquad -b_2 \qquad -b_1$

•——————•- - - - - - - - - - - - - - - - - - - -•——————•

Moreover the number of vertices of the graph (which is the number of components of the exceptional divisor) is odd, say $2h + 1$, and the lift of the involution to the resolution exchanges the extremal curves.

The exceptional divisor of the minimal resolution of the corresponding singular point of X has $h + 3$ component, all rational, and its dual graph is the following

Definition 4.3 We will say that such a singular point of X is of type $D_{n,q}$.

4.2 Invariants

Once we have computed all singular points of X, which means that we know exactly how many singular points of each type has our X, we can compute some of the characteristic numbers of the constructed quasi-étale surface S.

We give here the formulas [9, 16, 19] in terms of the construction data described at the beginning of this section, and of the singularities of X. Recall that the signatures of the generating vectors determine the genera g_i of the induced covers by the formula (2).

In the following, for each singular point of type $C_{n,q}$, we will denote by q' the only integer $0 < q' < n$ with $qq' \equiv 1 \bmod n$, and by $b_1, \ldots, b_k$ the coefficient of the continued fraction of $\frac{n}{q}$. Then $\frac{n}{q'} = [b_k, \ldots, b_1]$ and therefore a singularity of type $C_{n,q}$ is also of type $C_{n,q'}$.

For each singular point of type $C_{n,q}$ (e.g. [9])

$$k_x := -2 + \frac{2+q+q'}{n} + \sum_1^k (b_i - 2) \quad B_x := \frac{q+q'}{n} + \sum_1^k b_i$$

whereas, for singular points of type $D_{n,q}$ [16]

$$k_x := -1 + \frac{2+q+q'}{2n} + \sum_1^k \frac{b_i - 2}{2} \quad B_x := 6 + \frac{q+q'}{2n} + \sum_1^k \frac{b_i}{2}$$

All k_x, B_x are nonnegative. Then

$$K_S^2 = \frac{8(g_1 - 1)(g_2 - 1)}{|G|} - \sum_{x \in \mathrm{Sing}\, X} k_x$$

$$\chi(\mathcal{O}_S) = \frac{K_S^2}{8} + \frac{\sum_{x \in \mathrm{Sing}\, X} B_x}{24}$$

Finally, for the irregularity [27], in the unmixed case

$$q(S) = g'_1 + g'_2$$

whereas in the mixed case

$$q(S) = g'_1 = g'_2$$

The topological fundamental group, finally, can be computed directly by a method due to Armstrong [1, 2]. The computation is rather complicated, but can easily performed by a computer. For example, in [10] there is an implementation for that in the MAGMA [12] language.

So, if the surface is minimal ($S = \overline{S}$), we know all invariants of the constructed surfaces. As we will see in the next sections, this happens often, but not always, and determining K_S^2 is sometimes a challenging task.

5 The Minimality Problem

In the previous sections we have given a method to construct quasi-étale surfaces and to compute, from the construction of S, $\pi_1(S)$, K_S^2, $p_g(S)$ and $q(S)$. If S is minimal, then $K_{\tilde{S}}^2 = K_S^2$ and we can compute by Riemann-Roch Theorem all the plurigenera of S, and locate its position in the "geography".

This is often the case. Indeed K_X (which is the Weil divisor defined as closure of the canonical divisor of the open subset of the smooth points of X) pulls back to $K_{C_1 \times C_2}$ and therefore, if C_1 and C_2 have genus at least 2, it is big and nef. Therefore, if the action of G is free, then $X = S$ is minimal. On the other hand, if the action is not free, the singularities of X induce a discrepancy among K_S and the pull-back of K_X which is big and nef; if this discrepancy is big enough, the surface S may become non minimal or even not of general type. The first examples of non minimal unmixed quasi-étale surfaces of general type have been produced in [19], two surfaces with $p_g = q = 1$. The first example of an unmixed regular non minimal quasi-étale surfaces of general type has been produced in [9], a surface with $p_g = q = 0$; some more examples have been constructed in [7].

It is unclear if there exists a non minimal mixed quasi-étale surface of general type. Indeed we have the following

Theorem 5.1 ([16]) *Every irregular mixed quasi-étale surface of general type is minimal.*

The argument of the proof can't be extended to the regular case. Indeed [16] shows that, if S is a mixed quasi-étale surface of general type containing a smooth rational curve E with self intersection (-1), then the image of E on X pulls back to a curve of $C \times C$ whose image in $C' \times C'$ is a rational curve (here as before $C' := C/G^{(0)}$). Then, since the genus of C' equals the irregularity of the surface, we get a contradiction in the irregular case ($C' \times C'$ does not contain any rational curve), but nothing can be deduced in the regular case.

The unmixed irregular case, as (previously) shown in [19], is not much more complicated. Indeed if both curves C_1' and C_2' are irregular, then $C_1' \times C_2'$ does not contain any rational curve and one concludes as above (indeed this remark in [19] inspired the proof of Theorem 5.1). Else, up to exchanging C_1 and C_2, C_1' is rational, C_2' has genus $q > 0$ and all rational curves on S are contracted by the natural fibration $f_2 \colon S \to C_2'$. On the other hand, the only fibers of f_2 which are not isomorphic to C_1 are those whose image on X contains some singular points: it is not difficult then to compute explicitly their decomposition in irreducible components, and therefore describe all rational curves on S: in this way [19] could determine the minimal model of all the surfaces with $p_g = q = 1$ they constructed.

For example: [19] constructed one surface with $K_S^2 = 1$ whose quotient model had three singular points of respective type $\frac{1}{7}, \frac{2}{7}, \frac{2}{7}$ all in the image of the same fiber of f_2. That fiber depicted in Fig. 3, on the left, contains then all rational curves of S. It has six irreducible components, all smooth rational curves, 5 forming the exceptional divisor of the resolution of the singularities, and the sixth having self-intersection -1.

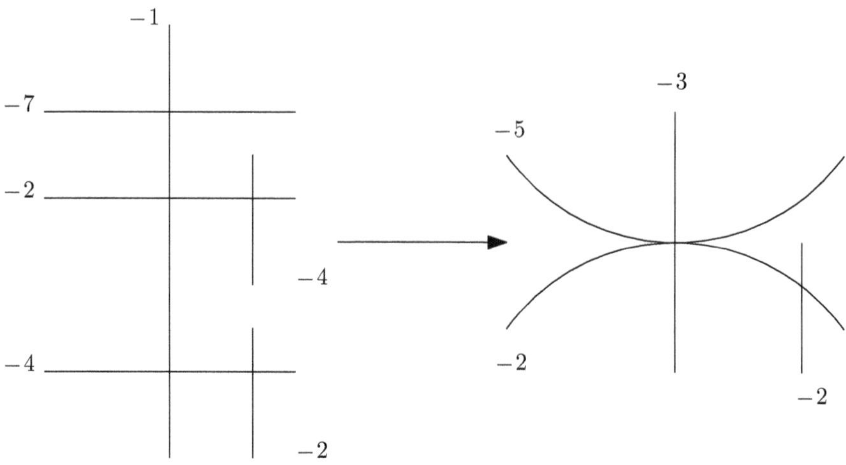

Fig. 3 The rational curves in the example by Polizzi and Mistretta

Contracting the (-1)-curve and the (-2)-curve transversal to it one gets a surface $\bar{S}$ with $K_{\bar{S}}^2 = 3$ with exactly 4 rational curves, depicted on the right of Fig. 3. The minimality follows since none has self intersection -1.

Determining the minimal model in the regular case may be much more challenging, since the rational curves are not forced to stay in the fibers of any of the two fibrations $f_i : S \to C_i/G$; indeed we are not able, in the regular case, to compute all rational curves on S. The surface constructed in [9], named "fake Godeaux" there, is a surface with $K_S^2 = 1$, $p_g = q = 0$ whose quotient model has exactly the same configuration of singularities, $\frac{1}{7}, \frac{2}{7}, \frac{2}{7}$, of the previous example. Bauer and Pignatelli [9] constructed two rational curves on S with self-intersection -1 intersecting the exceptional divisor of the resolution $S \to X$ as in Fig. 4, on the left. Contracting both we get a surface $\bar{S}$ with $K^2 = 3$, $p_g = q = 0$ with the funny configuration of rational curves on the right of Fig. 4.

Still, we know that ([9, Sect. 4] and [16, Corollary 4.7])

Proposition 5.2 *Let $\mathbb{P}^1 \to X$ be a rational curve on the quotient model of a quasi-étale surface such that C_1 and C_2 have both genus at least 2 (that's a necessary condition for X to be of general type). Then there are at least three distinct points of $\mathbb{P}^1$ mapped to singular points of X.*

This forces every further rational curve in $\bar{S}$ to intersect in at least three distinct points the configuration of curves on the right of Fig. 4. On the other hand ([16, Corollary 4.8])

Proposition 5.3 *Let E be a rational curve with self intersection -1 on a surface of general type, F be a reduced divisor whose support is made by rational curves of self-intersection either -2 or -3. Then $EF \leq 2$.*

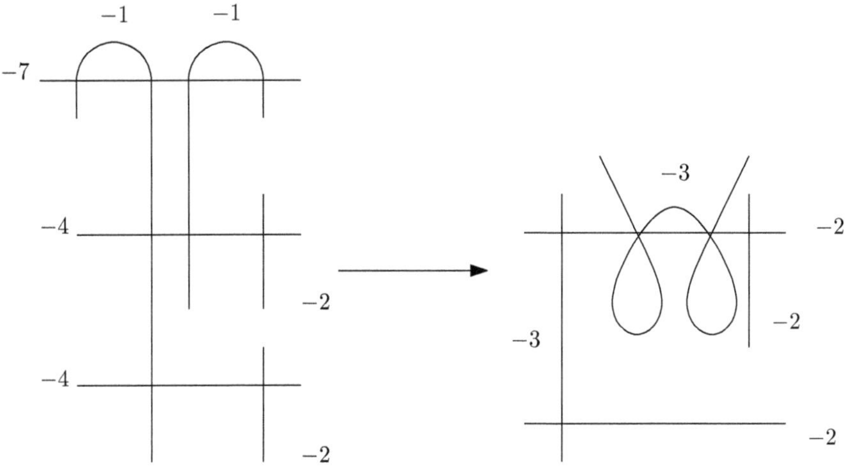

Fig. 4 The rational curves in the fake Godeaux surfaces

We know that $\overline{S}$ is of general type since by the Enriques-Kodaira classification a regular surface with $K^2 > 0$ is either of general type or rational, all rational surfaces are simply connected, and we can compute its fundamental group, which is $\mathbb{Z}_{/6\mathbb{Z}}$. Looking at the configuration of curves on the right of Fig. 4, by Propositions 5.2 and 5.3 we conclude that $\overline{S}$ is minimal.

A similar arguments works for proving the minimality of many of the surfaces constructed with this method, although one may need to substitute Proposition 5.3 with similar statements involving different configurations of rational curves. Some examples are in [9, Proposition 2.7].

We have not mentioned here how the two (-1)-curves on S are constructed; that's a rather complicated construction whose details goes beyond the scopes of this paper; the interested reader may read [9, Sect. 5]. We have recently found ([7]) few more examples of non minimal unmixed quasi-étale surfaces of general type with $p_g = q = 0$, and in all cases (up to now) we were able to determine their minimal model by constructing all the (-1)-curves on them. Still we have to run a different construction in each case: it would be nice to have a general method.

6 Surfaces of General Type with $\chi(\mathcal{O}_S) = 1$

This strategy has been used in the last year by several authors to construct new surfaces of general type, mainly for the "special" region of the geography of the surfaces with $\chi = 1$ (equivalently $p_g = q$). Thank to the contribution of, among others [4, 5, 9, 10, 13, 15, 16, 18–26, 28], we have now the complete list of all

Table 1 Minimal unmixed quasi-étale surfaces of general type with $p_g = q = 0$

Sing X	K_S^2	Sign.		G	#fam
	8	$2^3, 3$	5^3	$\langle 60, 5 \rangle$	1
	8	2^5	2^6	$\langle 8, 5 \rangle$	1
	8	2^5	2^5	$\langle 16, 14 \rangle$	1
	8	2^5	$3^2, 5$	$\langle 60, 5 \rangle$	1
	8	$2^2, 4^2$	$2^2, 4^2$	$\langle 16, 3 \rangle$	1
	8	3^4	3^4	$\langle 9, 2 \rangle$	1
	8	2^6	$3, 4^2$	$\langle 24, 12 \rangle$	1
	8	$2^2, 4^2$	$2^3, 4$	$\langle 32, 27 \rangle$	1
	8	$2, 5^2$	3^4	$\langle 60, 5 \rangle$	1
	8	$2, 4, 6$	2^6	$\langle 48, 48 \rangle$	1
	8	$2^3, 4$	2^6	$\langle 16, 11 \rangle$	1
	8	5^3	5^3	$\langle 25, 2 \rangle$	2
$\frac{1}{2}^2$	6	$2, 4, 6$	$2, 4, 10$	$\langle 240, 189 \rangle$	1
$\frac{1}{2}^2$	6	$2^3, 4$	$2^4, 4$	$\langle 16, 11 \rangle$	1
$\frac{1}{2}^2$	6	$2, 4, 6$	$2^4, 4$	$\langle 48, 48 \rangle$	1
$\frac{1}{2}^2$	6	$2, 3^3$	$2, 5^2$	$\langle 60, 5 \rangle$	1
$\frac{1}{2}^2$	6	$2, 7^2$	$3^2, 4$	$\langle 168, 42 \rangle$	2
$\frac{1}{2}^2$	6	$2, 5^2$	$3^2, 4$	$\langle 360, 118 \rangle$	2
$\frac{1}{2}^4$	4	$2^2, 3^2$	$2^2, 3^2$	$\langle 18, 4 \rangle$	1
$\frac{1}{2}^4$	4	$2^3, 4$	$2^3, 4$	$\langle 32, 27 \rangle$	1
$\frac{1}{2}^4$	4	$2^2, 4^2$	$2^2, 4^2$	$\langle 8, 2 \rangle$	1
$\frac{1}{2}^4$	4	2^5	2^5	$\langle 8, 5 \rangle$	1
$\frac{1}{2}^4$	4	$2, 4, 6$	2^5	$\langle 48, 48 \rangle$	1
$\frac{1}{2}^4$	4	$2, 5^2$	$2^2, 3^2$	$\langle 60, 5 \rangle$	1
$\frac{1}{2}^4$	4	2^5	$3, 4^2$	$\langle 24, 12 \rangle$	1
$\frac{1}{2}^4$	4	$2, 4, 6$	$2^2, 4^2$	$\langle 48, 48 \rangle$	1
$\frac{1}{2}^4$	4	$2^3, 4$	2^5	$\langle 16, 11 \rangle$	1
$\frac{1}{2}^4$	4	$2, 4, 5$	$3, 6^2$	$\langle 120, 34 \rangle$	1
$\frac{1}{2}^4$	4	$2^2, 3^2$	$3, 6^2$	$\langle 18, 3 \rangle$	1
$\frac{1}{2}^6$	2	$2^3, 4$	$2^3, 4$	$\langle 16, 11 \rangle$	1
$\frac{1}{2}^6$	2	$2, 4, 6$	$2^3, 4$	$\langle 48, 48 \rangle$	1
$\frac{1}{2}^6$	2	$2, 4, 5$	$2, 6^2$	$\langle 120, 34 \rangle$	1
$\frac{1}{2}^6$	2	$2, 3, 7$	4^3	$\langle 168, 42 \rangle$	2
$\frac{1}{2}^6$	2	4^3	4^3	$\langle 16, 2 \rangle$	1
$\frac{1}{2}^6$	2	$2, 5^2$	$2^3, 3$	$\langle 60, 5 \rangle$	1
$\frac{1}{2}^6$	2	$2, 6^2$	$2^3, 3$	$\langle 36, 10 \rangle$	1
$\frac{1}{3}, \frac{2}{3}$	5	$2^3, 3$	$3, 4^2$	$\langle 96, 227 \rangle$	1

(continued)

Table 1 (continued)

Sing X	K_S^2	Sign.		G	#fam
$\frac{1}{3}, \frac{2}{3}$	5	$2^4, 3$	$3, 4^2$	$\langle 24, 12 \rangle$	1
$\frac{1}{3}, \frac{2}{3}$	5	$2^3, 3$	$4^2, 6$	$\langle 48, 48 \rangle$	1
$\frac{1}{3}, \frac{2}{3}$	5	$2^3, 3$	$3, 5^2$	$\langle 60, 5 \rangle$	2
$\frac{1}{3}, \frac{2}{3}$	5	$2, 5, 6$	$3, 4^2$	$\langle 120, 34 \rangle$	1
$\frac{1}{3}, \frac{2}{3}$	5	$2, 4, 6$	$2^4, 3$	$\langle 48, 48 \rangle$	1
$\frac{1}{3}, \frac{1}{2}^2, \frac{2}{3}$	3	$2, 4, 6$	$2^2, 3, 4$	$\langle 48, 48 \rangle$	1
$\frac{1}{3}^2, \frac{2}{3}^2$	2	$2, 6^2$	$2^2, 3^2$	$\langle 24, 13 \rangle$	1
$\frac{1}{3}^2, \frac{2}{3}^2$	2	$3^2, 5$	$3^2, 5$	$\langle 75, 2 \rangle$	2
$\frac{1}{3}^2, \frac{2}{3}^2$	2	$2^3, 3$	$3^2, 5$	$\langle 60, 5 \rangle$	1
$\frac{1}{3}^2, \frac{2}{3}^2$	2	$2^2, 3^2$	$3, 4^2$	$\langle 24, 12 \rangle$	1
$\frac{1}{3}, \frac{1}{2}^4, \frac{2}{3}$	1	$2^3, 3$	$3, 4^2$	$\langle 24, 12 \rangle$	1
$\frac{1}{3}, \frac{1}{2}^4, \frac{2}{3}$	1	$2, 4, 6$	$2^3, 3$	$\langle 48, 48 \rangle$	1
$\frac{1}{3}, \frac{1}{2}^4, \frac{2}{3}$	1	$2, 3, 7$	$3, 4^2$	$\langle 168, 42 \rangle$	1
$\frac{1}{4}, \frac{1}{2}^2, \frac{3}{4}$	2	$2, 4, 7$	$3^2, 4$	$\langle 168, 42 \rangle$	2
$\frac{1}{4}, \frac{1}{2}^2, \frac{3}{4}$	2	$2, 4, 5$	$3^2, 4$	$\langle 360, 118 \rangle$	2
$\frac{1}{4}, \frac{1}{2}^2, \frac{3}{4}$	2	$2, 4, 5$	$3, 4, 6$	$\langle 120, 34 \rangle$	2
$\frac{2}{5}^2$	4	$2^3, 5$	$3^2, 5$	$\langle 60, 5 \rangle$	1
$\frac{2}{5}^2$	4	$2, 4, 5$	$3^2, 5$	$\langle 360, 118 \rangle$	1
$\frac{2}{5}^2$	4	$2, 4, 5$	$4^2, 5$	$\langle 160, 234 \rangle$	3
$\frac{1}{5}, \frac{4}{5}$	3	$2^3, 5$	$3^2, 5$	$\langle 60, 5 \rangle$	1
$\frac{1}{5}, \frac{4}{5}$	3	$2, 4, 5$	$3^2, 5$	$\langle 360, 118 \rangle$	1
$\frac{1}{5}, \frac{4}{5}$	3	$2, 4, 5$	$4^2, 5$	$\langle 160, 234 \rangle$	3

the minimal quasi-étale surfaces of general type with $\chi = 1$ and $K_S^2 > 0$, which includes all the minimal surfaces.

The surfaces of general type with $p_g = q \geq 3$ are classified (see [8] for a more precise account), so we only consider here the case $q \leq 2$. There are only three non minimal surfaces , which we have partially discussed in the previous section: the interested reader will find them in [9, 19].

The minimal quasi-étale surfaces of general type with $p_g = q \leq 2$ form few hundreds of families, whose construction is spread among several papers, sometimes using different notations. We take the opportunity offered by this paper to collect all of them in the same place with a coherent notation, reporting the full list in the Tables 1, 2, 3, 4, 5 and 6.

Table 2 Minimal mixed quasi-étale surfaces of general type with $p_g = q = 0$

Sing X	K_S^2	Sign	$G^{(0)}$	G	#fam
	8	2^5	$\langle 32, 46\rangle$	$\langle 64, 92\rangle$	1
	8	4^3	$\langle 128, 36\rangle$	$\langle 256, 3678\rangle$	3
	8	4^3	$\langle 128, 36\rangle$	$\langle 256, 3679\rangle$	1
$\frac{1}{2}^4$	4	2^5	$\langle 16, 11\rangle$	$\langle 32, 7\rangle$	1
$\frac{1}{2}^4$	4	2^5	$\langle 16, 14\rangle$	$\langle 32, 22\rangle$	1
$\frac{1}{2}^4$	4	4^3	$\langle 64, 23\rangle$	$\langle 128, 836\rangle$	1
$\frac{1}{2}^6$	2	2^5	$\langle 8, 5\rangle$	$\langle 16, 3\rangle$	1
$\frac{1}{2}^6$	2	4^3	$\langle 32, 2\rangle$	$\langle 64, 82\rangle$	1
$\frac{1}{3}^2, \frac{2}{3}^2$	2	$3^2, 4$	$\langle 384, 4\rangle$	$\langle 768, 1083540\rangle$	1
$\frac{1}{3}^2, \frac{2}{3}^2$	2	$3^2, 4$	$\langle 384, 4\rangle$	$\langle 768, 1083541\rangle$	1
$\frac{1}{2}^3, \frac{1}{4}^2$	2	$2^3, 4$	$\langle 64, 73\rangle$	$\langle 128, 1535\rangle$	1
$\frac{3}{8}, \frac{5}{8}$	3	$2^3, 8$	$\langle 32, 39\rangle$	$\langle 64, 42\rangle$	1
$\frac{1}{2}^2, D_{2,1}^2$	1	$2^3, 4$	$\langle 16, 11\rangle$	$\langle 32, 6\rangle$	1
$\frac{1}{2}^2, D_{2,1}^2$	2	$2^3, 4$	$\langle 32, 27\rangle$	$\langle 64, 32\rangle$	1
$\frac{1}{2}^2, D_{2,1}^2$	2	$2^2, 3^2$	$\langle 18, 4\rangle$	$\langle 36, 9\rangle$	1

A short explanation of the notation:

- The column Sing X gives the singularities of the quasi-étale quotient: we use the notation $\frac{q}{n}$ for the cyclic quotient singularities, $D_{n,q}$ for the singularities, in the mixed case, which are branching points of the double cover $Y \to X$. We use exponents for multiplicities: for example $\frac{1}{2}^2, D_{2,1}^2$ means that X has 4 singular points, 2 ordinary nodes and 2 of type $D_{2,1}$ (in particular Y has 6 nodes, and the fixed locus of the involution is given by two of them). In the case of surfaces isogenous of a product, equivalently if X is smooth, we leave this field blank.
- The column K_S^2 is self-explanatory.
- The columns Sign. give the involved signatures, two in the unmixed case, one in the mixed case. We use here, for short, exponents for representing the multiplicity, and omit g' when equal to zero. So $2^3, 3$ is a shortcut for $0; 2, 2, 2, 3$.
- The columns G and $G^{(0)}$ (the latter only in the mixed case) give the corresponding group in the MAGMA/GAP4 notation. So $\langle 60, 5\rangle$, for example, is the 5th group of order 60 in the MAGMA/GAP4 database of finite groups: this is the alternating group in 5 elements $\mathfrak{A}_5$.
- g_{alb} is the genus of the general fibre of the Albanese map, which is very important for the classification of the irregular surfaces. The column is missing for $q = 0$ since there is no Albanese map in that case. When, for $q = 2$, we leave it blank, it means that the Albanese map is not a fibration.
- In few cases there are 2 or 3 different families for which all the previous data coincide: instead of putting more identical rows, we used only one row for all of them, and add a last column, #fam, counting the number of families corresponding

Table 3 Minimal unmixed quasi-étale surfaces of general type with $p_g = q = 1$

Sing X	K_S^2	g_{alb}	Sign.		G	#fam
	8	3	2^6	$1; 2^2$	$\langle 4, 2 \rangle$	1
	8	3	2^5	$1; 2^2$	$\langle 8, 5 \rangle$	1
	8	3	$2^2, 4^2$	$1; 2^2$	$\langle 8, 2 \rangle$	2
	8	3	$2, 8^2$	$1; 2^2$	$\langle 16, 5 \rangle$	1
	8	3	$2^2, 4^2$	$1; 2^2$	$\langle 8, 3 \rangle$	1
	8	3	$2^3, 6$	$1; 2^2$	$\langle 12, 4 \rangle$	1
	8	3	$2^3, 4$	$1; 2^2$	$\langle 16, 11 \rangle$	1
	8	3	$2, 4, 12$	$1; 2^2$	$\langle 24, 5 \rangle$	1
	8	3	$2, 6^2$	$1; 2^2$	$\langle 24, 13 \rangle$	1
	8	3	$3, 4^2$	$1; 2^2$	$\langle 24, 12 \rangle$	1
	8	3	$2, 4, 8$	$1; 2^2$	$\langle 32, 9 \rangle$	1
	8	3	$2, 4, 6$	$1; 2^2$	$\langle 48, 48 \rangle$	1
	8	4	2^6	$1; 3$	$\langle 6, 1 \rangle$	1
	8	4	2^5	$1; 3$	$\langle 12, 4 \rangle$	1
	8	4	$2^2, 3^2$	$1; 3$	$\langle 18, 3 \rangle$	2
	8	4	$3, 6^2$	$1; 3$	$\langle 18, 3 \rangle$	1
	8	4	$2^3, 4$	$1; 3$	$\langle 24, 12 \rangle$	1
	8	4	$2, 6^2$	$1; 3$	$\langle 36, 10 \rangle$	1
	8	4	$2, 6^2$	$1; 3$	$\langle 36, 12 \rangle$	1
	8	4	$2, 4^2$	$1; 3$	$\langle 36, 9 \rangle$	2
	8	4	$2, 5^2$	$1; 3$	$\langle 60, 5 \rangle$	1
	8	4	$2, 3, 12$	$1; 3$	$\langle 72, 42 \rangle$	1
	8	4	$2, 4, 5$	$1; 3$	$\langle 120, 34 \rangle$	1
	8	5	2^6	$1; 2$	$\langle 8, 3 \rangle$	1
	8	5	3^4	$1; 2$	$\langle 12, 3 \rangle$	2
	8	5	$2^2, 4^2$	$1; 2$	$\langle 16, 3 \rangle$	3
	8	5	$2^2, 3^2$	$1; 2$	$\langle 24, 13 \rangle$	2
	8	5	$3, 6^2$	$1; 2$	$\langle 24, 13 \rangle$	1
	8	5	$2, 8^2$	$1; 2$	$\langle 32, 5 \rangle$	1
	8	5	$2, 8^2$	$1; 2$	$\langle 32, 7 \rangle$	1
	8	5	4^3	$1; 2$	$\langle 32, 2 \rangle$	1
	8	5	4^3	$1; 2$	$\langle 32, 6 \rangle$	1
	8	5	$2, 6^2$	$1; 2$	$\langle 48, 49 \rangle$	1
	8	5	$2, 4, 8$	$1; 2$	$\langle 64, 32 \rangle$	2
	8	5	$2, 5^2$	$1; 2$	$\langle 80, 49 \rangle$	2
$\frac{1}{2}^2$	6	3	$2, 5^2$	$1; 2$	$\langle 24, 3 \rangle$	
$\frac{1}{2}^2$	6	3	$2, 5^2$	$1; 2$	$\langle 32, 9 \rangle$	
$\frac{1}{2}^2$	6	3	$2, 5^2$	$1; 2$	$\langle 32, 11 \rangle$	
$\frac{1}{2}^2$	6	3	$2, 5^2$	$1; 2$	$\langle 48, 33 \rangle$	

(continued)

Table 3 (continued)

Sing X	K_S^2	g_{alb}	Sign.		G	#fam
$\frac{1}{2}^2$	6	3	2^6	$1; 2^2$	$\langle 48, 3\rangle$	
$\frac{1}{2}^2$	6	3	2^5	$1; 2^2$	$\langle 168, 42\rangle$	
$\frac{1}{2}^2$	6	4	$2^2, 4^2$	$1; 2^2$	$\langle 8, 3\rangle$	
$\frac{1}{2}^2$	6	4	$2, 8^2$	$1; 2^2$	$\langle 12, 3\rangle$	
$\frac{1}{2}^2$	6	4	$2^2, 4^2$	$1; 2^2$	$\langle 24, 10\rangle$	
$\frac{1}{2}^2$	6	4	$2^3, 6$	$1; 2^2$	$\langle 36, 11\rangle$	
$\frac{1}{2}^2$	6	4	$2^3, 4$	$1; 2^2$	$\langle 72, 40\rangle$	
$\frac{1}{2}^2$	6	4	$2, 4, 12$	$1; 2^2$	$\langle 120, 34\rangle$	
$\frac{1}{3}, \frac{2}{3}$	5	3	$2, 6^2$	$1; 2^2$	$\langle 6, 1\rangle$	
$\frac{1}{3}, \frac{2}{3}$	5	3	$3, 4^2$	$1; 2^2$	$\langle 12, 1\rangle$	
$\frac{1}{3}, \frac{2}{3}$	5	3	$2, 4, 8$	$1; 2^2$	$\langle 12, 4\rangle$	
$\frac{1}{3}, \frac{2}{3}$	5	3	$2, 4, 6$	$1; 2^2$	$\langle 24, 5\rangle$	
$\frac{1}{3}, \frac{2}{3}$	5	3	2^6	$1; 3$	$\langle 24, 12\rangle$	
$\frac{1}{3}, \frac{2}{3}$	5	3	2^5	$1; 3$	$\langle 48, 48\rangle$	
$\frac{1}{3}, \frac{2}{3}$	5	3	$2^2, 3^2$	$1; 3$	$\langle 96, 64\rangle$	
$\frac{1}{3}, \frac{2}{3}$	5	3	$3, 6^2$	$1; 3$	$\langle 168, 42\rangle$	
$\frac{1}{2}^4$	4	2	$2^3, 4$	$1; 3$	$\langle 4, 2\rangle$	
$\frac{1}{2}^4$	4	2	$2, 6^2$	$1; 3$	$\langle 6, 2\rangle$	
$\frac{1}{2}^4$	4	2	$2, 6^2$	$1; 3$	$\langle 6, 1\rangle$	
$\frac{1}{2}^4$	4	2	$2, 4^2$	$1; 3$	$\langle 8, 3\rangle$	
$\frac{1}{2}^4$	4	2	$2, 5^2$	$1; 3$	$\langle 12, 5\rangle$	
$\frac{1}{2}^4$	4	2	$2, 3, 12$	$1; 3$	$\langle 12, 4\rangle$	
$\frac{1}{2}^4$	4	2	$2, 4, 5$	$1; 3$	$\langle 16, 8\rangle$	
$\frac{1}{2}^4$	4	2	2^6	$1; 2$	$\langle 24, 8\rangle$	
$\frac{1}{2}^4$	4	2	3^4	$1; 2$	$\langle 48, 29\rangle$	
$\frac{1}{2}^4$	4	3	$2^2, 4^2$	$1; 2$	$\langle 8, 3\rangle$	
$\frac{1}{2}^4$	4	3	$2^2, 3^2$	$1; 2$	$\langle 12, 3\rangle$	
$\frac{1}{2}^4$	4	3	$3, 6^2$	$1; 2$	$\langle 16, 6\rangle$	
$\frac{1}{2}^4$	4	3	$3, 6^2$	$1; 2$	$\langle 16, 4\rangle$	
$\frac{1}{2}^4$	4	3	$2, 8^2$	$1; 2$	$\langle 24, 13\rangle$	
$\frac{1}{2}^2, \frac{1}{3}, \frac{2}{3}$	3	2	$2, 8^2$	$1; 2$	$\langle 24, 8\rangle$	
$\frac{1}{2}^2, \frac{1}{3}, \frac{2}{3}$	3	2	4^3	$1; 2$	$\langle 48, 29\rangle$	
$\frac{1}{2}^2, \frac{1}{4}, \frac{3}{4}$	2	2	4^3	$1; 2$	$\langle 16, 8\rangle$	
$\frac{1}{2}^2, \frac{1}{4}, \frac{3}{4}$	2	2	4^3	$1; 2$	$\langle 24, 3\rangle$	
$\frac{1}{3}^2, \frac{2}{3}^2$	2	2	$2, 6^2$	$1; 2$	$\langle 6, 1\rangle$	

(continued)

Table 3 (continued)

Sing X	K_S^2	g_{alb}	Sign.		G	#fam
$\frac{1}{3}^2, \frac{2}{3}^2$	2	2	2, 4, 8	1; 2	$\langle 12, 1 \rangle$	
$\frac{1}{3}^2, \frac{2}{3}^2$	2	2	2, 5^2	1; 2	$\langle 12, 4 \rangle$	
$\frac{1}{2}^6$	2	2	2, 5^2	1; 2	$\langle 8, 4 \rangle$	
$\frac{1}{2}^6$	2	2	2, 5^2	1; 2	$\langle 8, 3 \rangle$	

Table 4 Minimal mixed quasi-étale surfaces of general type with $p_g = q = 1$

Sing X	K_S^2	g_{alb}	Sign	$G^{(0)}$	G	#fam
	8	5	1; 2^2	$\langle 8, 2 \rangle$	$\langle 16, 6 \rangle$	1
	8	5	1; 2^2	$\langle 8, 3 \rangle$	$\langle 16, 8 \rangle$	1
	8	5	1; 2^2	$\langle 8, 5 \rangle$	$\langle 16, 3 \rangle$	1
$\frac{1}{2}^2$	6	3	1;2	$\langle 24, 13 \rangle$	$\langle 48, 30 \rangle$	1
$\frac{1}{2}^2$	6	7	1;2	$\langle 24, 13 \rangle$	$\langle 48, 31 \rangle$	1
$\frac{1}{2}^4$	4	3	1; 2^2	$\langle 4, 1 \rangle$	$\langle 8, 1 \rangle$	1
$\frac{1}{2}^4$	4	3	1; 2^2	$\langle 4, 2 \rangle$	$\langle 8, 2 \rangle$	1
$\frac{1}{2}^4$	4	2	1;2	$\langle 16, 3 \rangle$	$\langle 32, 29 \rangle$	1
$\frac{1}{2}^4$	4	3	1;2	$\langle 16, 4 \rangle$	$\langle 32, 13 \rangle$	1
$\frac{1}{2}^4$	4	3	1;2	$\langle 16, 4 \rangle$	$\langle 32, 14 \rangle$	1
$\frac{1}{2}^4$	4	2	1;2	$\langle 16, 4 \rangle$	$\langle 32, 32 \rangle$	1
$\frac{1}{2}^4$	4	2	1;2	$\langle 16, 4 \rangle$	$\langle 32, 35 \rangle$	1
$\frac{1}{2}^4$	4	3	1;2	$\langle 16, 5 \rangle$	$\langle 32, 15 \rangle$	1
$\frac{1}{3}, \frac{2}{3}$	5	3	1;3	$\langle 12, 1 \rangle$	$\langle 24, 4 \rangle$	1
$\frac{1}{3}, \frac{2}{3}$	5	3	1;3	$\langle 12, 4 \rangle$	$\langle 24, 5 \rangle$	1
$\frac{3}{5}$	6	5	1;5	$\langle 10, 1 \rangle$	$\langle 20, 3 \rangle$	1
$\frac{1}{2}, D_{2,1}^2$	2	2	1; 2^2	$\langle 2, 1 \rangle$	$\langle 4, 1 \rangle$	1
$\frac{1}{2}, D_{2,1}^2$	2	2	1;2	$\langle 8, 3 \rangle$	$\langle 16, 8 \rangle$	1
$\frac{1}{2}, D_{2,1}^2$	2	2	1;2	$\langle 8, 4 \rangle$	$\langle 16, 9 \rangle$	1

to the row. The reader will not find the column #fam in the latter part of Table 3; indeed in that cases the number of families corresponding to each row is not known, at least to the author.

It is worth noticing that all possible values of K_S^2 in the yellow region of Fig. 2 are obtained already by the Table 1, so by surfaces with $p_g = q = 0$. In the irregular case, since an inequality of Debarre shows that for a minimal irregular surface of general type $K_S^2 \geq 2 p_g$, also the quasi-étale surfaces with $p_g = q = 1$ and $p_g = q = 2$

Table 5 Minimal unmixed quasi-étale surfaces of general type with $p_g = q = 2$

Sing X	K_S^2	g_{alb}	Sign.		G	#fam
	8	2	2^6	2;	$\langle 2, 1 \rangle$	1
	8	2	3^4	2;	$\langle 3, 1 \rangle$	1
	8	2	2^5	2;	$\langle 4, 2 \rangle$	2
	8	2	$2^2, 4^2$	2;	$\langle 4, 1 \rangle$	1
	8	2	5^3	2;	$\langle 5, 1 \rangle$	1
	8	2	$2^2, 3^2$	2;	$\langle 6, 2 \rangle$	1
	8	2	$3, 6^2$	2;	$\langle 6, 2 \rangle$	1
	8	2	$2, 8^2$	2;	$\langle 8, 1 \rangle$	1
	8	2	$2, 5, 10$	2;	$\langle 10, 2 \rangle$	1
	8	2	$2, 6^2$	2;	$\langle 12, 5 \rangle$	2
	8	2	$2^2, 3^2$	2;	$\langle 6, 1 \rangle$	1
	8	2	4^3	2;	$\langle 8, 4 \rangle$	1
	8	2	$2^3, 4$	2;	$\langle 8, 3 \rangle$	2
	8	2	$2^3, 3$	2;	$\langle 12, 4 \rangle$	2
	8	2	$3, 4^2$	2;	$\langle 12, 1 \rangle$	1
	8	2	$2, 4, 8$	2;	$\langle 16, 8 \rangle$	1
	8	2	$2, 4, 6$	2;	$\langle 24, 8 \rangle$	2
	8	2	$3^2, 4$	2;	$\langle 24, 3 \rangle$	1
	8	2	$2, 3, 8$	2;	$\langle 48, 29 \rangle$	1
	8		$1; 2^2$	$1; 2^2$	$\langle 4, 2 \rangle$	1
	8		$1; 3$	$1; 2^2$	$\langle 6, 1 \rangle$	1
	8		$1; 2$	$1; 2^2$	$\langle 8, 3 \rangle$	1
$\frac{1}{2}^2$	6		$1; 2$	$1; 2$	$\langle 12, 3 \rangle$	1
$\frac{1}{2}^4$	4		$1; 2^2$	$1; 2^2$	$\langle 2, 1 \rangle$	1
$\frac{1}{2}^4$	4		$1; 2$	$1; 2$	$\langle 8, 3 \rangle$	1
$\frac{1}{2}^4$	4		$1; 2$	$1; 2$	$\langle 8, 4 \rangle$	1
$\frac{1}{3}, \frac{2}{3}$	5		$1; 3$	$1; 3$	$\langle 6, 1 \rangle$	1

Table 6 Minimal mixed quasi-étale surfaces of general type with $p_g = q = 2$

Sing X	K_S^2	Sign	$G^{(0)}$	G	#fam
	8	2;	$\langle 2, 1 \rangle$	$\langle 4, 1 \rangle$	1

reach all values of K_S^2 attained by minimal surfaces of general type: this supports our claim that this method could possibly fill the yellow region.

It would be nice to have a complete classification of all unmixed quasi-étale surfaces of general type with $\chi(\mathcal{O}_S) = 1$; the list is complete only in the mixed irregular case by Theorem 5.1.

The methods developed in the mentioned papers allow in principle to construct the whole list of quasi-étale surfaces with a fixed value of the triple (p_g, q, K_S^2). Since when we blow up a point the value of K_S^2 drops by 1, there is no lower bound for the value of K_S^2 of a surface of general type in terms of birational invariants as p_g and q; in particular these methods can't give a complete classification unless one can prove such a lower bound for quasi-étale surfaces; at the moment we do not have even a reasonable conjecture for that bound.

A different method, which could possibly in the future produce a complete classification of the quasi-étale surfaces with $\chi = 1$, has been recently presented in [7], where K_S^2 is substituted by a different number related to the structure of the Néron-Severi group of a quasi-étale surface.

References

1. M.A. Armstrong, On the fundamental group of an orbit space. Proc. Camb. Philos. Soc. **61**, 639–646 (1965)
2. M.A. Armstrong, The fundamental group of the orbit space of a discontinuous group. Proc. Camb. Philos. Soc. **64**, 299–301 (1968)
3. W. Barth, C. Peters, A. van de Ven, *Compact Complex Surfaces*. Ergebnisse der Mathematik und ihrer Grenzgebiete.3, vol. 4 (Springer, Berlin, 1984), pp. x+304
4. I. Bauer, F. Catanese, Some new surfaces with $p_g = q = 0$. *The Fano Conference* (University of Torino, Turin, 2004), pp. 123–142
5. I. Bauer, F. Catanese, F. Grunewald, The classification of surfaces with $p_g = q = 0$ isogenous to a product of curves. Pure Appl. Math. Q. **4**(2), part 1, 547–586 (2008)
6. I. Bauer, F. Catanese, R. Pignatelli, Surfaces of general type with geometric genus zero: a survey. *Complex and Differential Geometry*. Springer Proceedings in Mathematics, vol. 8 (2011), pp. 1–48
7. I. Bauer, R. Pignatelli, Product-Quotient Surfaces: new invariants and algorithms. To appear on Groups, Geometry and Dynamics. arxiv:math/1308.5508
8. I. Bauer, F. Catanese, R. Pignatelli, Complex surfaces of general type: some recent progress. *Global Aspects of Complex Geometry* (Springer, Berlin, 2006), pp. 1–58
9. I. Bauer, R. Pignatelli, The classification of minimal product-quotient surfaces with $p_g = 0$. Math. Comput. **81**(280), 2389–2418 (2012)
10. I. Bauer, F. Catanese, F. Grunewald, R. Pignatelli, Quotients of a product of curves by a finite group and their fundamental groups. Am. J. Math. **134**(4), 993–1049 (2012)
11. A. Beauville, *Complex Algebraic Surfaces*. Translated from the French by R. Barlow, N.I. Shepherd-Barron, M. Reid. London Mathematical Society Lecture Note Series, vol. 68 (Cambridge University Press, Cambridge, 1983). pp. iv+132
12. W. Bosma, J. Cannon, C. Playoust, The Magma algebra system. I: the user language. J. Symb. Comput. **24**(3–4), 235–265 (1997)
13. G. Carnovale, F. Polizzi, The classification of surfaces with $p_g = q = 1$ isogenous to a product of curves. Adv. Geom. **9**, 233–256 (2009)

14. F. Catanese, Fibred Surfaces, varieties isogeneous to a product and related moduli spaces. Am. J. Math. **122**(1), 1–44 (2000)
15. D. Frapporti, Mixed surfaces, new surfaces of general type with $p_g = 0$ and their fundamental group. To appear on Coll. Math. arXiv:1105.1259
16. D. Frapporti, R. Pignatelli, Mixed quasi-étale quotients with arbitrary singularities. Glasgow Math. J. **57**, 143–165 (2015)
17. A. Garbagnati, M. Penegini, K3 surfaces with a non-symplectic automorphism and product-quotient surfaces with cyclic groups. arXiv:1303.1653
18. M. Mendes Lopes, R. Pardini, The bicanonical map of surfaces with $p_g = 0$ and $K^2 \geq 7$. Bull. Lond. Math. Soc. **33**, 265–274 (2001)
19. E. Mistretta, F. Polizzi, Standard isotrivial fibrations with $p_g = q = 1$ II. J. Pure Appl. Algebra **214**, 344–369 (2010)
20. R. Pardini, The classification of double planes of general type with $K^2 = 8$ and $p_g = 0$. J. Algebra **259**, 95–118 (2003)
21. M. Penegini, The classification of isotrivially fibred surfaces with $p_g = q = 2$, and topics on beauville surfaces. Ph.D Thesis University of Bayreuth (2010)
22. M. Penegini, The classification of isotrivially fibred surfaces with $p_g = q = 2$. With an appendix by S. Rollenske. Collect. Math. **62**, 239–274 (2011)
23. F. Polizzi, Surfaces of general type with $p_g = q = 1$, $K^2 = 8$ and bicanonical map of degree 2. Trans. Am. Math. Soc. **358**, 759–798 (2006)
24. F. Polizzi, On surfaces of general type with $p_g = q = 1$ isogenous to a product of curves. Commun. Algebra **36**, 2023–2053 (2008)
25. F. Polizzi, Standard isotrivial fibrations with $p_g = q = 1$. J. Algebra **321**, 1600–1631 (2009)
26. F. Polizzi, Numerical properties of isotrivial fibrations. Geom. Dedicata **147**, 323–355 (2010)
27. F. Serrano, Isotrivial fibred surfaces. Ann. Mat. Pura Appl. **171**(4), 63–81 (1996)
28. F. Zucconi, Surfaces with $p_g = q = 2$ and an irrational pencil. Can. J. Math. **55**, 649–672 (2003)

Isotrivially Fibred Surfaces and Their Numerical Invariants

Francesco Polizzi

Abstract We give a survey of our previous work on relatively minimal isotrivial fibrations $\alpha\colon X \longrightarrow C$, where X is a smooth, projective surface and C is a curve. In particular, we consider two inequalities involving the numerical invariants K_X^2 and $\chi(\mathcal{O}_X)$ and we illustrate them by means of several examples and counter examples.

2010 Mathematics Subject Classification 14J99, 14J29

1 Introduction

One of the most useful tools in the study of algebraic surfaces is the analysis of *fibrations*, that is morphisms with connected fibres from a smooth surface X to a curve C. A fibration $\alpha\colon X \longrightarrow C$ is called *isotrivial* when all its smooth fibres are isomorphic; a deep investigation of such kind of fibrations can be found in [6].

A special kind of isotrivial fibrations are the standard ones, whose definition is as follows. A smooth surface X is called a *standard isotrivial fibration* if there exists a finite group G, acting faithfully on two smooth curves C_1 and C_2 and diagonally on their product, such that X is isomorphic to the minimal desingularization of $T = (C_1 \times C_2)/G$. Then X has two isotrivial fibrations $\alpha_i\colon X \longrightarrow C_i/G$, induced by the natural projections $p_i\colon C_1 \times C_2 \longrightarrow C_i$.

If the action of G on $C_1 \times C_2$ is free, we say that $X = T$ is a *quasi-bundle*. For instance, unmixed Beauville surfaces are defined as rigid quasi-bundles, i.e., quasi-bundles such that $C_i/G \cong \mathbb{P}^1$ and the covers $C_i \longrightarrow \mathbb{P}^1$ are both branched at three points. Therefore the classification of Beauville surfaces can be reduced to combinatorial finite group theory involving triangle groups; we refer the reader to the other papers in this Volume for further details.

Standard isotrivial fibrations were thoroughly investigated by Serrano in [16, 17]; in particular he showed, by a monodromy argument, that every isotrivial fibration

F. Polizzi (✉)
Dipartimento di Matematica e Informatica, Università della Calabria,
Cubo 30B, 87036 Arcavacata Rende, Cosenza, Italy
e-mail: polizzi@mat.unical.it

© Springer International Publishing Switzerland 2015
I. Bauer et al. (eds.), *Beauville Surfaces and Groups*, Springer Proceedings
in Mathematics & Statistics 123, DOI 10.1007/978-3-319-13862-6_11

is isomorphic to a standard one. Since then, such fibrations have been widely used in order to produce new examples of minimal surfaces of general type with small birational invariants, in particular with $p_g = q = 0$ [1–5], with $p_g = q = 1$ [7, 10, 12–15] and with $p_g = q = 2$ [11].

In this paper we discuss the following theorem, that was obtained in [15]. Let $f\colon X \longrightarrow C$ be any relatively minimal, isotrivial fibration with $g(C) \geq 1$. If X is neither ruled nor isomorphic to a quasi-bundle, then $K_X^2 \leq 8\chi(\mathcal{O}_X) - 2$. If, in addition, K_X is ample, then $K_X^2 \leq 8\chi(\mathcal{O}_X) - 5$. This generalizes previous results of Serrano and Tan [17, 18].

This work is organized as follows. In Sect. 2 we set up the notation and the terminology, we state our theorem and we provide a sketch of its proof. In Sect. 3 we exhibit several examples and counterexamples illustrating its meaning. More precisely, Examples 3.1 and 3.2 imply that both the above inequalities involving K_X^2 and $\chi(\mathcal{O}_X)$ are sharp, whereas Examples 3.3 and 3.4 show that, when K_X is not ample, both cases $K_X^2 = 8\chi(\mathcal{O}_X) - 3$ and $K_X^2 = 8\chi(\mathcal{O}_X) - 4$ actually occur. Finally, in Example 3.5 we describe an isotrivially fibred surface X with $K_X^2 = 8\chi(\mathcal{O}_X) - 5$ and K_X *not* ample. In all these examples X is a minimal surface of general type with $p_g = q = 1$, obtained as a standard isotrivial fibration.

Notations and conventions. All varieties in this article are defined over $\mathbb{C}$. If X is a projective, non-singular surface X then K_X denotes the canonical class, $p_g(X) = h^0(X, K_X)$ is the *geometric genus*, $q(S) = h^1(X, K_X)$ is the *irregularity* and $\chi(\mathcal{O}_X) = 1 - q(X) + p_g(X)$ is the *Euler characteristic*.

If G is a finite group and $g \in G$, we denote by $|G|$ and $o(g)$ the orders of G and g, respectively.

2 The Main Result

Definition 2.1 Let X be a smooth, complex projective surface and let $\alpha\colon X \longrightarrow C$ be a fibration onto a smooth curve C. We say that α is *isotrivial* if all its smooth fibres are isomorphic.

Definition 2.2 A smooth surface S is called a *standard isotrivial fibration* if there exists a finite group G, acting faithfully on two smooth projective curves C_1 and C_2 and diagonally on their product, so that S is isomorphic to the minimal desingularization of $T := (C_1 \times C_2)/G$. We denote such a desingularization by $\lambda\colon S \longrightarrow T$.

If $\lambda\colon S \longrightarrow T = (C_1 \times C_2)/G$ is any standard isotrivial fibration, composing the two projections $\pi_1\colon T \longrightarrow C_1/G$ and $\pi_2\colon T \longrightarrow C_2/G$ with λ one obtains two morphisms $\alpha_1\colon S \longrightarrow C_1/G$ and $\alpha_2\colon S \longrightarrow C_2/G$, whose smooth fibres are isomorphic to C_2 and C_1, respectively. One also has $q(S) = g(C_1/G) + g(C_2/G)$, see [9].

A monodromy argument ([17, Sect. 2]) implies that any isotrivial fibration $\alpha\colon$ $X \longrightarrow C$ is birational to a standard one; in other words, there exists $T = (C_1 \times C_2)/G$ and a birational map $T \dashrightarrow X$ such that the following diagram

$$\begin{array}{ccc}
S & & \\
\downarrow{\scriptstyle\lambda} & \searrow{\scriptstyle\psi} & \\
T & \dashrightarrow & X \\
\downarrow{\scriptstyle\pi_2} & & \downarrow{\scriptstyle\alpha} \\
C_2/G & \overset{\cong}{\longrightarrow} & C
\end{array} \qquad (1)$$

commutes.

When the action of G is free, then $S = T$ is called a *quasi-bundle*; in this case one has $K_S^2 = 8\chi(\mathcal{O}_S)$. In 1996, F. Serrano and, indipendently, S. L. Tan improved this result, showing that for any isotrivial fibration $\alpha\colon X \longrightarrow C$ one has

$$K_X^2 \leq 8\chi(\mathcal{O}_X) \qquad (2)$$

and that the equality holds if and only if X is either ruled or isomorphic to a quasi bundle [17, 18]. Serrano's proof is based on a fine analysis of the projective bundle $\mathbb{P}(\Omega_X^1)$, whereas Tan's proof uses base change techniques.

This paper deals with the following refinement of (2), that we proved in [15].

Theorem 2.3 *Let $\alpha\colon X \longrightarrow C$ be any relatively minimal isotrivial fibration, with $g(C) \geq 1$. If X is neither ruled nor isomorphic to a quasi bundle, then*

$$K_X^2 \leq 8\chi(\mathcal{O}_X) - 2, \qquad (3)$$

and if the equality holds then X is a minimal surface of general type whose canonical model has precisely two ordinary double points as singularities. Moreover, under the further assumption that K_X is ample, we have

$$K_X^2 \leq 8\chi(\mathcal{O}_X) - 5. \qquad (4)$$

Finally, inequalities 3 and 4 are sharp.

Let us give a sketch of the proof of Theorem 2.3, whose full details can be found in [15]. Let us consider a standard isotrivial fibration S birational to X and the corresponding commutative diagram (1). Since α is relatively minimal and $g(C) \geq 1$, the surface X is a minimal model. Moreover we are assuming that X is not ruled, so K_X is nef and the birational map $\psi\colon S \longrightarrow X$ is actually a morphism [8, Proposition 8], which induces an isomorphism of X with the minimal model S_m of S. Since C_2/G has positive genus, all the (-1)-curves of S are necessarily contained in fibres of $\alpha_2\colon S \longrightarrow C_2/G$, hence our isotrivial fibration $\alpha\colon X \longrightarrow C$ is equivalent to an isotrivial fibration $\alpha_m\colon S_m \longrightarrow C_2/G$.

The surface T contains at most a finite number of isolated singularities that, locally analytically, look like the quotient of $\mathbb{C}^2$ by the action of the cyclic group $\mathbb{Z}/n\mathbb{Z} = \langle \xi \rangle$ defined by $\xi \cdot (x, y) = (\xi x, \xi^q y)$, where $0 < q < n$, $(n, q) = 1$ and ξ is a primitive n-th root of unity. We call this singularity a *cyclic quotient singularity* of type $\frac{1}{n}(1, q)$. The exceptional divisor $\mathcal{E}$ of its minimal resolution is a HJ-string (abbreviation of Hirzebruch-Jung string), that is to say, a connected union $\mathcal{E} = \bigcup_{i=1}^{k} Z_i$ of smooth rational curves $Z_1, \ldots, Z_k$ with self-intersection $-b_i := Z_i^2 \leq -2$, and ordered linearly so that $Z_i Z_{i+1} = 1$ for all i, and $Z_i Z_j = 0$ if $|i - j| \geq 2$. More precisely, given the continued fraction

$$\frac{n}{q} = [b_1, \ldots, b_k] = b_1 - \cfrac{1}{b_2 - \cfrac{1}{\cdots - \cfrac{1}{b_k}}}, \quad b_i \geq 2,$$

the dual graph of $\mathcal{E}$ is

$$\overset{-b_1}{\bullet} \!-\!\!-\!\! \overset{-b_2}{\bullet} \!-\!-\!- \overset{-b_{k-1}}{\bullet} \!-\!\!-\!\! \overset{-b_k}{\bullet}$$

For instance, the cyclic quotient singularities $\frac{1}{n}(1, n-1)$ are precisely the rational double points of type A_{n-1}; in particular, the singularities $\frac{1}{2}(1, 1)$ are the ordinary double points.

Finally, the invariants $K_{S_m}^2$ and $e(S_m)$ can be computed knowing the number and type of the singularities of T. In fact, since $g(C_2/G) = g(C) \geq 1$, all the (-1)-curves of S are components of reducible fibres of $\alpha_2 \colon S \longrightarrow C_2/G$; it follows that it is possible to define, for any such a reducible fibre F, an invariant $\delta(F) \in \mathbb{Q}$ such that

$$K_{S_m}^2 = 8\chi(\mathcal{O}_{S_m}) - \sum_{F \text{ reducible}} \delta(F). \tag{5}$$

The proof of Theorem 2.3 follows from a careful analysis of the possible values of $\delta(F)$, based on some identities on continued fractions which are a consequence of the so-called Riemenschneider's duality between the HJ-expansions of $\frac{n}{q}$ and $\frac{n}{n-q}$.

3 Examples

Let us denote by $\Gamma(0 \mid m_1, \ldots, m_r)$ the abstract group of Fuchsian type with presentation

$$\left\langle g_1, \ldots, g_r \mid g_k^{m_k} = 1, \ \prod_{i=1}^{r} g_i = 1 \right\rangle$$

and by $\Gamma(1 \mid n_1, \ldots, n_s)$ the abstract group of Fuchsian type with presentation

$$\left\langle l_1, \ldots, l_s, \ h_1, h_2 \mid l_k^{n_k} = 1, \ \ [h_1, h_2] \prod_{i=1}^{s} l_i = 1 \right\rangle.$$

We call the sets $(m_1, \ldots, m_r)$ and $(n_1, \ldots, n_s)$ the *branching data* of these groups. For convenience we make abbreviations such as $(2^3, 3^2)$ for $(2, 2, 2, 3, 3)$ when we write down the branching data.

Example 3.1 This example appears in [14, Sect. 7].

Take $G = D_8$, the dihedral group of order 8, with presentation

$$G = \langle x, y \mid x^2 = y^4 = 1, xy = y^3 x \rangle.$$

There are two epimorphisms of groups

$$\varphi \colon \Gamma(0|2^4, 4) \longrightarrow G, \quad \psi \colon \Gamma(1|2) \longrightarrow G$$

defined in the following way:

$$\varphi(g_1) = x, \quad \varphi(g_2) = xy, \quad \varphi(g_3) = x, \quad \psi(g_4) = xy^2, \quad \psi(g_5) = y,$$
$$\psi(\ell_1) = y^2, \quad \psi(h_1) = y, \quad \psi(h_2) = x. \tag{6}$$

By Riemann Existence Theorem [10, Proposition 1.3] they induce two G-coverings

$$f_1 \colon C_1 \longrightarrow \mathbb{P}^1 \cong C_1/G, \quad f_2 \colon C_2 \longrightarrow E \cong C_2/G,$$

where E is an elliptic curve, $g(C_1) = 4$ and $g(C_2) = 3$. Moreover, f_1 is branched at five points with branching orders 2, 2, 2, 2, 4, whereas f_2 is branched at one point with branching order 2.

By (6) it follows that the unique element of G, different from the identity, which acts with fixed points on both C_1 and C_2 is y^2. Denoting by $|\text{Fix}_{C_i}(g)|$ the number of fixed points of $g \in G$ on the curve C_i, we obtain

$$|\text{Fix}_{C_1}(y^2)| = 2, \quad |\text{Fix}_{C_2}(y^2)| = 4,$$

so we have $4 \cdot 2 = 8$ points in $C_1 \times C_2$ whose stabilizer is non-trivial (it is isomorphic to $\langle y^2 \rangle$). The G-orbit of each of them has cardinality $|G|/o(y^2) = 4$, hence we have exactly two singular points in $T = (C_1 \times C_2)/G$. More precisely, since $\langle y^2 \rangle$ has order 2, it follows

$$\text{Sing}\, T = 2 \times \frac{1}{2}(1, 1).$$

Let $\lambda \colon X \longrightarrow T$ be the minimal resolution of singularities of T; then X is a surface of general type, whose numerical invariants can be computed by using [10, Proposition 5.1]; we obtain

$$p_g(X) = q(X) = 1, \quad K_X^2 = 6.$$

Fig. 1 The unique singular
fibre of $\alpha\colon X \longrightarrow E$ in
Examples 3.1 and 3.4

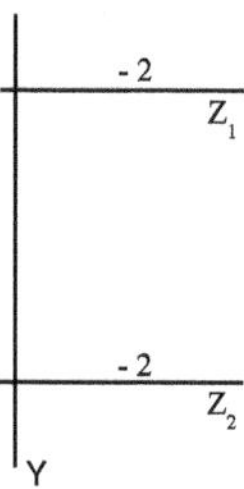

The isotrivial fibration $\pi_2\colon T \longrightarrow C_2/G \cong E$ yields, after composition with λ, an
isotrivial fibration $\alpha\colon X \longrightarrow E$, which is the Albanese morphism of X. The fibration
α has a unique singular fibre $F = 2Y + Z_1 + Z_2$, where the Z_i are two disjoint (-2)
curves and $YZ_i = 1$, see Fig. 1.

Since E is an elliptic curve, any (-1)-curve of X must be a component of F. On
the other hand, we have $K_X F = 2g(C_1) - 2 = 6$ and $F^2 = 0$, hence $K_X Y = 3$ and
$Y^2 = -1$. Thus Y is not a (-1)-curve, so X is a minimal surface of general type
satisfying

$$K_X^2 = 8\chi(\mathcal{O}_X) - 2$$

and T is the canonical model of X. This example shows that inequality (3) in Theorem 2.3 is sharp. Notice that K_X is *not* ample, since $K_X Z_i = 0$.

Example 3.2 This example can be found in [10, Sect. 5].

Take as G the semi-direct product $\mathbb{Z}/3\mathbb{Z} \ltimes (\mathbb{Z}/4\mathbb{Z})^2$ whose presentation is

$$G = \langle x, y, z \mid x^3 = y^4 = z^4 = 1, [y, z] = 1, xyx^{-1} = z, xzx^{-1} = (yz)^{-1}\rangle.$$

There are two epimorphisms of groups

$$\varphi\colon \Gamma(0|3^2, 4) \longrightarrow G, \quad \psi\colon \Gamma(1|4) \longrightarrow G$$

defined in the following way:

$$\varphi(g_1) = x, \quad \varphi(g_2) = x^2 y^3, \quad \varphi(g_3) = y,$$
$$\psi(\ell_1) = y, \quad \psi(h_1) = x, \quad \psi(h_2) = xyxy^2. \tag{7}$$

By Riemann Existence Theorem they induce two G-coverings

$$f_1\colon C_1 \longrightarrow \mathbb{P}^1 \cong C_1/G, \quad f_2\colon C_2 \longrightarrow E \cong C_2/G,$$

where E is an elliptic curve, $g(C_1) = 3$ and $g(C_2) = 19$. Moreover, f_1 is branched
at three points with branching orders 3, 3, 4, whereas f_2 is branched at one point
with branching order 4.

By (7) it follows that the nontrivial elements of G having fixed points on $C_1 \times C_2$ are precisely those in the set

$$\Sigma = \bigcup_{\sigma \in G} \langle \sigma y \sigma^{-1} \rangle \setminus \{1\}$$

and the elements of order 4 in Σ are $\{y, z, y^3 z^3, y^3, z^3, yz\}$. The product surface $C_1 \times C_2$ contains exactly 48 points with nontrivial stabilizer, and for each of them the G-orbit has cardinality $|G|/o(y) = 12$, thus $T = (C_1 \times C_2)/G$ contains 4 singular points.

Moreover the conjugacy class of y in G is $\{y, z, y^3 z^3\}$, whereas the conjugacy class of y^3 is $\{y^3, z^3, yz\}$. Therefore, for any $h \in \Sigma$ with $o(h) = 4$, one has that h is not conjugate to h^{-1} in G. Looking at the local action of G around each of the fixed points, this implies that

$$\text{Sing } T = 4 \times \frac{1}{4}(1, 1).$$

Let $\lambda \colon X \longrightarrow T$ be the minimal resolution of singularities of X; then X is a surface of general type whose invariants are

$$p_g(X) = q(X) = 1, \quad K_X^2 = 2.$$

The isotrivial fibration $\pi_2 \colon T \longrightarrow C_2/G \cong E$ yields, after composition with λ, an isotrivial fibration $\alpha \colon X \longrightarrow E$ which is the Albanese morphism of X. The isotrivial fibration α has a unique singular fibre $F = 4Y + A_1 + A_2 + A_3 + A_4$, where the A_i are disjoint smooth rational curves such that $A_i^2 = -4$ and $Y A_i = 1$ (see Fig. 2).

We have $K_X F = 2g(C_1) - 2 = 4$ and $F^2 = 0$, so we deduce $K_X Y = Y^2 = -1$. Thus Y is a (-1)-curve, which is necessarily the unique (-1)-curve in X. Since Y is contained in a fibre of $\alpha \colon X \longrightarrow E$, after blowing down Y we obtain another isotrivial fibration $\alpha_m \colon X_m \longrightarrow E$ such that

$$p_g(X_m) = q(X_m) = 1, \quad K_{X_m}^2 = 3,$$

that is

$$K_{X_m}^2 = 8\chi(\mathcal{O}_{X_m}) - 5. \tag{8}$$

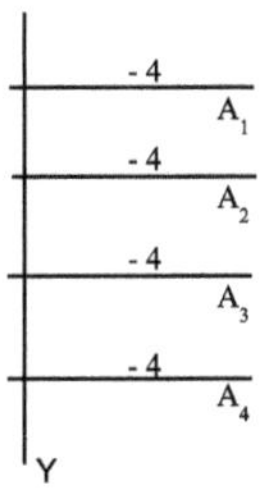

Fig. 2 The unique singular fibre of $\alpha \colon X \longrightarrow E$ in Example 3.2

Fig. 3 The unique singular
fibre of $\alpha_m : X_m \longrightarrow E$ in
Example 3.2

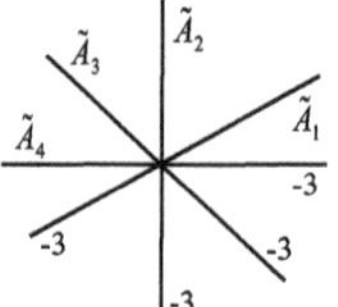

The isotrivial fibration α_m contains a unique singular fibre, namely the image of F in X_m, that is illustrated in Fig. 3.

Here each $\widetilde{A}_i$ is a smooth rational curve with self-intersection -3; in particular X_m contains neither (-1)-curves nor (-2)-curves. This shows that X_m is a minimal model and that K_{X_m} is ample, so (8) implies that inequality (4) in Theorem 2.3 is sharp.

The next two examples show that, if K_X is not ample, then both cases $K_X^2 = 8\chi(\mathcal{O}_X) - 3$ and $K_X^2 = 8\chi(\mathcal{O}_X) - 4$ may actually occur.

Example 3.3 We exhibit an isotrivial fibration with K_X not ample and $K_X^2 = 8\chi(\mathcal{O}_X) - 3$; this example appears in [10, Sect. 5].

Take $G = D_{12}$, the finite dihedral group of order 12, with presentation

$$G = \langle x, y | x^2 = y^6 = 1, xy = y^5 x \rangle.$$

There are two epimorphisms of groups

$$\varphi \colon \Gamma(0|2^3, 6) \longrightarrow G, \quad \psi \colon \Gamma(1|3) \longrightarrow G$$

defined in the following way:

$$\varphi(g_1) = x, \quad \varphi(g_2) = xy^2, \quad \varphi(g_3) = y^3, \quad \psi(g_4) = y,$$
$$\psi(\ell_1) = y^2, \quad \psi(h_1) = x, \quad \psi(h_2) = y. \tag{9}$$

By Riemann Existence Theorem they induce two G-coverings

$$f_1 \colon C_1 \longrightarrow \mathbb{P}^1 \cong C_1/G, \quad f_2 \colon C_2 \longrightarrow E \cong C_2/G,$$

where E is an elliptic curve, $g(C_1) = 3$ and $g(C_2) = 5$. Moreover, f_1 is branched at four points with branching orders 2, 2, 2, 6, whereas f_2 is branched at one point with branching order 3. By (9) it follows that the nontrivial elements of G having fixed points on $C_1 \times C_2$ are precisely those in the set

$$\Sigma = \bigcup_{\sigma \in G} \langle \sigma y^2 \sigma^{-1} \rangle \setminus \{1\} = \{y^2, y^4\}.$$

The product surface $C_1 \times C_2$ contains exactly 8 points with nontrivial stabilizer, and for each of them the G-orbit has cardinality $|G|/o(y^2) = 4$, thus $T = (C_1 \times C_2)/G$ contains two singular points. Since y^2 is conjugate to y^4 in G, looking at the local action of G around each of the fixed points one obtains

$$\operatorname{Sing} T = \frac{1}{3}(1, 1) + \frac{1}{3}(1, 2).$$

Let $\lambda \colon X \longrightarrow T$ be the minimal resolution of singularities of X; then X is a surface of general type whose invariants are

$$p_g(X) = q(X) = 1, \quad K_X^2 = 5,$$

that is

$$K_X^2 = 8\chi(\mathcal{O}_X) - 3.$$

The isotrivial fibration $\pi_2 \colon T \longrightarrow C_2/G \cong E$ yields, after composition with λ, an isotrivial fibration $\alpha \colon X \longrightarrow E$ which is the Albanese morphism of X. The isotrivial fibration α has a unique singular fibre $F = 3Y + A + 2B_1 + B_2$, where A is a (-3)-curve and the B_i are (-2)-curves, see Fig. 4.

Using $K_X F = 2g(C_1) - 2 = 4$ and $F^2 = 0$ one obtains $K_X Y = 1$ and $Y^2 = -1$, hence Y is not a (-1)-curve and X is a minimal model. Notice that K_X is not ample, since X contains the (-2)-curves B_i.

Example 3.4 We exhibit an isotrivial fibration with K_X not ample and $K_X^2 = 8\chi(\mathcal{O}_X) - 4$; this example appears in [14, Sect. 6].

Take $G = \mathbb{Z}/2\mathbb{Z} \times \mathbb{Z}/2\mathbb{Z}$, with presentation

$$G = \langle x, y \,|\, x^2 = y^2 = 1, [x, y] = 1 \rangle.$$

There are two epimorphisms of groups

$$\varphi \colon \Gamma(0|2^5) \longrightarrow G, \quad \psi \colon \Gamma(1|2^2) \longrightarrow G$$

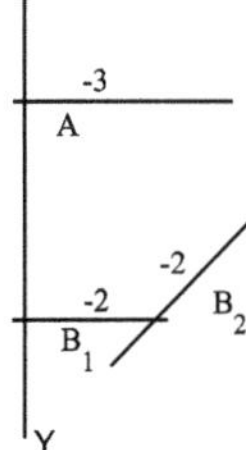

Fig. 4 The unique singular fibre of $\alpha \colon X \longrightarrow E$ in Example 3.3

defined in the following way:

$$\varphi(g_1) = x, \quad \varphi(g_2) = y, \quad \varphi(g_3) = xy, \quad \psi(g_4) = xy, \quad \varphi(g_5) = xy$$
$$\psi(\ell_1) = x, \quad \psi(\ell_2) = x, \quad \psi(h_1) = y, \quad \psi(h_2) = y. \tag{10}$$

By Riemann Existence Theorem they induce two G-coverings

$$f_1 \colon C_1 \longrightarrow \mathbb{P}^1 \cong C_1/G, \quad f_2 \colon C_2 \longrightarrow E \cong C_2/G,$$

where E is an elliptic curve, $g(C_1) = 2$ and $g(C_2) = 3$. Moreover, f_1 is branched at five points, f_2 is branched at two points and all the branching orders are equal to 2. By (10) it follows that the unique nontrivial element of G having fixed points on $C_1 \times C_2$ is x. The product surface $C_1 \times C_2$ contains exactly 8 points with nontrivial stabilizer, and for each of them the G-orbit has cardinality $|G|/o(x) = 2$, thus $T = (C_1 \times C_2)/G$ contains four singular points. More precisely, since $\langle x \rangle$ has order 2, we have

$$\mathrm{Sing}\, T = 4 \times \frac{1}{2}(1, 1).$$

Let $\lambda \colon X \longrightarrow T$ be the minimal resolution of singularities of X; then X is a surface of general type whose invariants are

$$p_g(X) = q(X) = 1, \quad K_X^2 = 4.$$

The isotrivial fibration $\pi_2 \colon T \longrightarrow C_2/G \cong E$ yields, after composition with λ, an isotrivial fibration $\alpha \colon X \longrightarrow E$ which is the Albanese morphism of X. The isotrivial fibration α has two singular fibres F_1 and F_2, which look like the singular fibre in Fig. 1, i.e. they are of the form $F_i = 2Y_i + Z_{1i} + Z_{2i}$, where the Z_{ji} are disjoint (-2)-curves. Using $2 = K_X F_i = 2K_X Y_i$ and $F_i^2 = 0$ one obtains $K_X Y_i = 1$ and $Y_i^2 = -1$. In particular Y_i is not a (-1)-curve, so X is a minimal surface of general type satisfying

$$K_X^2 = 8\chi(\mathcal{O}_X) - 4.$$

Notice that K_X is not ample, since X contains the four (-2)-curves Z_{ji}.

Looking at Theorem 2.3, one might ask whether any isotrivially fibred surface X with $K_X^2 = 8\chi(\mathcal{O}_X) - 5$ has ample canonical class. The answer is negative, as shown by the following example that can be found in [10, Sect. 6]

Example 3.5 Take $G = D_{3,7,2}$, namely the metacyclic group of order 21 whose presentation is
$$G = \langle x, y \mid x^3 = y^7 = 1, xyx^{-1} = y^2 \rangle.$$

There are two epimorphisms of groups

$$\varphi \colon \Gamma(0|3^2, 7) \longrightarrow G, \quad \psi \colon \Gamma(1|7) \longrightarrow G$$

defined in the following way:

$$\varphi(g_1) = x^2, \quad \varphi(g_2) = xy^6, \quad \varphi(g_3) = y,$$
$$\psi(\ell_1) = y, \quad \psi(h_1) = y, \quad \psi(h_2) = x. \tag{11}$$

By Riemann Existence Theorem they induce two G-coverings

$$f_1 \colon C_1 \longrightarrow \mathbb{P}^1 \cong C_1/G, \quad f_2 \colon C_2 \longrightarrow E \cong C_2/G,$$

where E is an elliptic curve, $g(C_1) = 3$ and $g(C_2) = 10$. Moreover, f_1 is branched at three points with branching orders 3, 3, 7, whereas f_2 is branched at one point with branching order 7.

By (11) it follows that the nontrivial elements of G having fixed points on $C_1 \times C_2$ are precisely those in the set

$$\Sigma = \bigcup_{\sigma \in G} \langle \sigma y \sigma^{-1} \rangle \setminus \{1\} = \{y, y^2, y^3, y^4, y^5, y^6\}.$$

The product surface $C_1 \times C_2$ contains exactly 9 points with non-trivial stabilizer, and for each of them the G-orbit has cardinality $|G|/o(y) = 3$; thus $T = (C_1 \times C_2)/G$ contains 3 singular points.

Moreover, the conjugacy class of y in G is $\{y, y^2, y^4\}$ and the conjugacy class of y^3 is $\{y^3, y^5, y^6\}$. In particular, every element in Σ is conjugate to its inverse. Looking at the local action of G around each of the fixed points, this implies that

$$\mathrm{Sing}\, T = \frac{1}{7}(1, 1) + \frac{1}{7}(1, 2) + \frac{1}{7}(1, 4).$$

Let $\lambda \colon X \longrightarrow T$ be the minimal resolution of singularities of X; then X is a surface of general type, whose invariants are

$$p_g(X) = q(X) = 1, \quad K_X^2 = 1.$$

The isotrivial fibration $\pi_2 \colon T \longrightarrow C_2/G \cong E$ yields, after composition with λ, an isotrivial fibration $\alpha \colon X \longrightarrow E$, which is the Albanese morphism of X. The fibration α has a unique singular fibre $F = 7Y + 4A_1 + A_2 + 2B_1 + B_2 + C$, which looks as in Fig. 5.

Here A_i, B_i and C are smooth rational curves, and the integer over each of them denotes as usual the corresponding self-intersection.

Using $K_X F = 4$ and $F^2 = 0$ we obtain $K_X Y = Y^2 = -1$, hence Y is the unique (-1)-curve in X. The minimal model X_m of X is obtained by first contracting Y and then the image of A_1. Since Y is contained in a fibre of $\alpha \colon X \longrightarrow E$, we obtain another isotrivial fibration $\alpha_m \colon X_m \longrightarrow E$ such that

$$p_g(X_m) = q(X_m) = 1, \quad K_{X_m}^2 = 3, \tag{12}$$

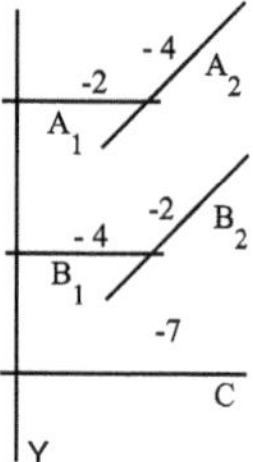

Fig. 5 The unique singular fibre of $\alpha\colon X \longrightarrow E$ in Example 3.5

that is

$$K^2_{X_m} = 8\chi(\mathcal{O}_{X_m}) - 5.$$

However, the canonical class K_{X_m} is *not* ample, as X_m contains two (-2)-curves, namely the images of B_1 and B_2.

Acknowledgments This paper is an expanded version of the talk given by the author at the conference *Beauville surfaces and Groups*, Newcastle University (UK), 7–9th June 2012. The author is grateful to the organizers N. Barker, I. Bauer, S. Garion and A. Vdovina for the invitation and the kind hospitality. He was partially supported by Progetto MIUR di Rilevante Interesse Nazionale *Geometria delle Varietà Algebriche e loro Spazi di Moduli*. He also thanks the referee, whose comments helped to improve the presentation of these results.

References

1. I. Bauer, F. Catanese, Some new surfaces with $p_g = q = 0$, in *Proceedings of the Fano Conference 123142* (University of Turin, Torino, 2004)
2. I. Bauer, F. Catanese, F. Grunewald, The classification of surfaces with $p_g = q = 0$ isogenous to a product of curves. Pure Appl. Math. Q. **4**(2), part 1, 547–5861 (2008)
3. I. Bauer, F. Catanese, R. Pignatelli, Surfaces of general type with geometric genus zero: a survey. *Complex and Differential Geometry*, Springer Proceedings in Mathematics, vol. 8, pp. 1–48 (2011)
4. I. Bauer, F. Catanese, F. Grunewald, R. Pignatelli, Quotient of a product of curves by a finite group and their fundamental groups. Am. J. Math. **134**(4), 993–1049 (2012)
5. I. Bauer, R. Pignatelli, The classification of minimal product-quotient surfaces with $p_g = 0$. Math. Comput. **81**(280), 2389–2418 (2012)
6. F. Catanese, Fibred surfaces, varieties isogenous to a product and related moduli spaces. Am. J. Math. **122**, 1–44 (2000)
7. G. Carnovale, F. Polizzi, The classification of surfaces of general type with $p_g = q = 1$ isogenous to a product. Adv. Geom. **9**, 233–256 (2009)
8. Classification of complex algebraic surfaces from the point of view of Mori theory (2004). http://www-fourier.ujf-grenoble.fr/~peters/surface.f/surf-spec.pdf
9. E. Freitag, Uber die Struktur der Funktionenkörper zu hyperabelschen Gruppen I. J. Reine. Angew. Math. **247**, 97–117 (1971)
10. E. Mistretta, F. Polizzi, Standard isotrivial fibrations with $p_g = q = 1$. *II*. J. Pure Appl. Algebra **214**, 344–369 (2010)
11. M. Penegini, The classification of isotrivially fibred surfaces with $p_g = q = 2$ (with an appendix by Sonke Rollenske). Collect. Math. **62**(3), 239–274 (2011)

12. F. Polizzi, Surfaces of general type with $p_g = q = 1$, $K_S^2 = 8$ and bicanonical map of degree 2. Trans. Am. Math. Soc. **358**(2), 759–798 (2006)
13. F. Polizzi, On surfaces of general type with $p_g = q = 1$ isogenous to a product of curves. Commun. Algebra **36**, 2023–2053 (2008)
14. F. Polizzi, Standard isotrivial fibrations with $p_g = q = 1$. J. Algebra **321**, 1600–1631 (2009)
15. F. Polizzi, Numerical properties of isotrivial fibrations. Geom. Dedicata **147**, 323–355 (2010)
16. F. Serrano, Fibrations on algebraic surfaces, in *Geometry of Complex Projective Varieties (Cetraro 1990)*, ed. by A. Lanteri, M. Palleschi, D.C. Struppa. Mediterranean Press, pp. 291–300 (1993)
17. F. Serrano, Isotrivial fibred surfaces. Annali di Matematica pura e applicata **CLXXI**, 63–81 (1996)
18. S.L. Tan, On the invariant of base changes of pencils of curves, II. Math. Z. **222**, 655–676 (1996)
19. The GAP Group, GAP—Groups, Algorithms, and Programming, Version 4.4 (2006). http://www.gap-system.org

GPSR Compliance
The European Union's (EU) General Product Safety Regulation (GPSR) is a set
of rules that requires consumer products to be safe and our obligations to
ensure this.

If you have any concerns about our products, you can contact us on

ProductSafety@springernature.com

In case Publisher is established outside the EU, the EU authorized
representative is:

Springer Nature Customer Service Center GmbH
Europaplatz 3
69115 Heidelberg, Germany

www.ingramcontent.com/pod-product-compliance
Ingram Content Group UK Ltd.
Pitfield, Milton Keynes, MK11 3LW, UK
UKHW021907280526
471597UK00003B/28